GW01157290

The Trillion-Dollar Stake of Carbon Neutrality

Energy Infrastructures in Hong Kong and the Greater Bay Area

The Trillion-Dollar Stake of Carbon Neutrality

Energy Infrastructures in Hong Kong and the Greater Bay Area

Sam Liao Cooke
Yuan Xu

The Chinese University of Hong Kong, Hong Kong

World Scientific

NEW JERSEY • LONDON • SINGAPORE • BEIJING • SHANGHAI • HONG KONG • TAIPEI • CHENNAI • TOKYO

Published by

World Scientific Publishing Co. Pte. Ltd.
5 Toh Tuck Link, Singapore 596224
USA office: 27 Warren Street, Suite 401-402, Hackensack, NJ 07601
UK office: 57 Shelton Street, Covent Garden, London WC2H 9HE

Library of Congress Cataloging-in-Publication Data
Names: Cooke, Sam Liao, author. | Xu, Yuan, author.
Title: The trillion-dollar stake of carbon neutrality : energy infrastructures in Hong Kong and the Greater Bay Area / Sam Liao Cooke, Yuan Xu (The Chinese University of Hong Kong, Hong Kong).
Description: New Jersey : World Scientific, [2024] | Includes bibliographical references.
Identifiers: LCCN 2024007317 | ISBN 9789811290305 (hardcover) | ISBN 9789811290312 (ebook) | ISBN 9789811290329 (ebook other)
Subjects: LCSH: Energy industries--China--Hong Kong. | Energy industries--China--Macau (Special Administrative Region) | Carbon dioxide mitigation--Economic aspects--China--Hong Kong. | Carbon dioxide mitigation--Economic aspects--China--Macau (Special Administrative Region) | Energy policy--Economic aspects--China--Hong Kong. | Energy policy--Economic aspects--China--Macau (Special Administrative Region)
Classification: LCC HD9502.C62 C66 2024 | DDC 333.790951--dc23/eng/20240321
LC record available at https://lccn.loc.gov/2024007317

British Library Cataloguing-in-Publication Data
A catalogue record for this book is available from the British Library.

Cover illustration by Antao Song and Anlan Xu.

Copyright © 2024 by World Scientific Publishing Co. Pte. Ltd.

All rights reserved. This book, or parts thereof, may not be reproduced in any form or by any means, electronic or mechanical, including photocopying, recording or any information storage and retrieval system now known or to be invented, without written permission from the publisher.

For photocopying of material in this volume, please pay a copying fee through the Copyright Clearance Center, Inc., 222 Rosewood Drive, Danvers, MA 01923, USA. In this case permission to photocopy is not required from the publisher.

For any available supplementary material, please visit
https://www.worldscientific.com/worldscibooks/10.1142/13762#t=suppl

Desk Editors: Soundararajan Raghuraman/Amanda Yun

Typeset by Stallion Press
Email: enquiries@stallionpress.com

Preface

Climate change is the most expensive environmental crisis that has ever been encountered by humankind. Its key cause is the emissions of greenhouse gases, especially CO_2 from fossil fuel consumption. The Intergovernmental Panel on Climate Change (IPCC) in 2018 illustrated the urgency of reducing CO_2 emissions to zero as early as 2040 to control the global temperature increase to below 1.5°C. Climate change incurs huge economic and social damages, and its mitigation competes for financial resources that may be better spent on other important issues. Nevertheless, hidden behind the crisis, economic opportunities are emerging and should be actively explored.

Climate mitigation creates new investment opportunities and new industries. Renewable energy industries have been thriving in the past decade to encroach on the share of fossil fuels, accelerate energy and economic transitions, and reshape the economic landscape. New jobs are flourishing. Coal played a crucial role in the industrial revolution, but as the most CO_2-intensive fuel, its consumption has been experiencing chilling headwinds, and coal jobs are declining. Global oil and gas sectors mainly feature mature corporations from developed countries, whereas wind and solar manufacturing have witnessed many more young firms from emerging countries, especially China. Electric vehicles are posing steadily growing threats to oil-based ones and exerting prospectively transformative impacts on conventional vehicle manufacturing around the globe. The more leveled competition encourages the potential reshuffling of international and domestic vehicle markets, for example, to enable the rapid rise of Tesla. Although China has been the largest auto market in the world since 2009, most new oil-based vehicles are still foreign brands. However, the country now accounts for more than half of global sales, mostly with domestic brands. With the accelerating shift toward electric vehicles, the industry could be dramatically transformed in the coming decades.

Prosperous new industries provide economic opportunities for ambitious firms and new start-ups but do not necessarily guarantee their individual success. Market competition in these new industries tends to be fiercer. The demise of the mighty Standard Oil in 1911 was not driven by market competition; rather, it resulted from regulatory forces meant to break its monopoly. By restricting access to scarce fossil-fuel resources, existing firms can effectively raise market entry barriers to constrain potential competitors. However, wind and solar energy resources are geographically much more widely distributed to significantly minimize such barriers. Innovation is becoming even more important in these fast-evolving industries that feature more firms, breakneck competition, and vibrant entry and exit. In 2013, China's Suntech, the largest solar PV manufacturer in the world not long before, went bankrupt. The largest wind turbine manufacturers in China have also experienced rapid reshuffling in the past 15 years.

Wind and solar PV industries and jobs are geographically more mobile for countries to compete for. Trade for fossil fuels mainly indicates the import and export of physically extracted natural resources, while the availability of fossil-fuel resources greatly constrains the location of fossil-fuel production and jobs. In contrast, wind turbines and solar PV panels can be manufactured in any country and easily shipped globally. Their economic production and jobs are thus much less geographically bound, although this situation is also more likely to lead to trade tensions. The United States in 2012 and the European Union in 2013 imposed punitive import tariffs on Chinese solar PV panels. A grand-scale trade war broke out between the United States and China in 2018 without a clear ending date in sight. The rapid rise of renewable energy and the prospective decline of fossil fuels will redistribute jobs across countries and probably result in more trade disputes. Furthermore, amidst the intensified competition in innovations, a crucial focus is on innovators who are highly mobile across countries. In the past decade, China has substantially increased funding for research and development to push up indigenous innovation and has been gradually reversing a brain drain, with more resources available for attracting established and young talents from abroad. However, such competition has escalated disputes over intellectual property rights.

Poverty alleviation also receives a helping hand from renewable energy development. The world still had nearly one billion people without access to electricity in 2017. The widely distributed renewable energy and their small-scale systems could help reach remote households and communities that are economically more expensive and technologically more challenging to be connected to electric grids. In addition, regions with rich solar energy resources, such as a desert, should have much sunshine and less rainfall. Those with rich wind energy resources should have strong wind. Especially when they are not fortunate enough to have other abundant natural resources, these unfavorable conditions could constrain economic development and thus lead to higher poverty rates. However, climate mitigation essentially elevates the economic value of their wind and solar energy resources. Their large-scale utilization and inter-regional electricity transmission could bring in investment, economic opportunities, and jobs to the poverty-trapped or economically less developed communities.

Climate mitigation may shake up the current global financial system, which has petrodollars as its key foundation. Global trade and financial transactions are dominated by US dollars, and US supremacy is built significantly upon this foundation. Oil is the single largest traded commodity across country borders. The oil embargo led by Saudi Arabia against the United States and several other countries in October 1973 created a major oil crisis and sent the developed world into economic recessions. However, in 1974, the two countries reached an agreement that Saudi Arabia would accept and only accept US dollars when exporting their oil and the petrodollars would be invested in US treasuries. Many other oil-exporting countries followed suit, resulting in the petrodollar system. Although the United States has become the largest oil producer in the world and China has grown to be the largest oil importer, the central role of US dollars has remained unchanged. If climate mitigation eventually undermines oil consumption and trade, leading to less demand for US dollars, the petrodollar system might start shaking from its core, casting uncertainties over future global finance and economics.

The Paris Agreement in 2015 successfully built a united, although not perfect, front for combating climate change. After swift ratification by many countries, it formally entered into force in November 2016. Although concerns about economic burden triggered the withdrawal of the United States in 2017, an economy should not turn a blind eye to the associated economic opportunities and impacts. An economy's pursuit of economic self-interests in climate mitigation, either domestically or internationally, could contribute to global efforts.

Downscaling from the megatrends, this book joins the debates by focusing on Hong Kong and the Greater Bay Area, which has 86 million residents and an economy of US$1.7 trillion, together with other less developed municipalities in Guangdong Province. China announced in September 2020 its goal to achieve carbon neutrality by 2060. As one of the most developed and carbon-intensive regions in China, the Greater Bay Area is supposed to bear more responsibility, with a very likely advanced timeline. Hong Kong has already committed to reaching the goal by 2050. We document and examine existing and planned individual energy infrastructure projects, mainly involving fossil fuels, renewables (wind and solar), nuclear, energy storage, CO_2 capture and storage (CCS), and others. Gaps are evaluated, and the necessary pace of changes in the next three decades is sketched. The financial stake for the region is up to the level of trillion US dollars. Stranded assets could be serious obstacles, while economic opportunities are also at the trillion US dollar scale to fundamentally reshape the region's energy economic landscape.

This book introduces readers to an in-depth understanding of carbon neutrality from the perspective of energy infrastructures. Those interested in carbon neutrality, climate change, infrastructure planning, energy businesses, or environmental, social and governance (ESG) will find the book helpful.

About the Authors

Sam Liao Cooke is a sustainability consultant at The Lantau Group and a former research assistant in the Department of Geography and Resource Management at the Chinese University of Hong Kong. He is also the former CEO of CarbonX, an environmental B2B startup based in Hong Kong that helped environmental startups increase revenue and find seed investments. Previously, he worked for a local NGO called OceansAsia, the Asian Development Bank, and Peking University. His research has been cited widely in mainstream media globally, including the *BBC*, *ABC*, *NBC*, *New York Times*, *South China Morning Post*, *British Medical Journal*, and Asian Development Bank reports. In his spare time, Sam is a scuba diving instructor and sailor, and when not working, he can often be found ether on or under water. He holds a BS degree in environmental studies from Eckerd College and an MSc in ecological economics from the University of Edinburgh and Scotland's Rural College (SRUC).

Yuan Xu is an associate professor in the Department of Geography and Resource Management and leads the Environmental Policy and Governance Program in the Institute of Environment, Energy and Sustainability at the Chinese University of Hong Kong. He is currently serving as the President of the Professional Association for China's Environment (PACE). He has published widely on energy and environmental policy, governance and strategy, and technological and industrial development, covering fossil fuels and renewables. His recent monographs include *Environmental Policy and Air Pollution in China: Governance and Strategy* and *Residential Electricity Consumption in Urbanizing China: Time Use and Climate-Friendly Living* (co-authored), both published by Routledge. Before joining CUHK in August 2010, he received a PhD degree in public policy from the Woodrow Wilson School of Public and International Affairs, Princeton University, and was a postdoctoral research associate in the Industrial Performance Center, Massachusetts Institute of Technology. He also holds an MS degree in climatology, a BS degree in atmospheric sciences, and a bachelor's degree in economics, all from Peking University.

Acknowledgments

Sam Cooke thanks his parents, Fang Li Cooke and Glenn Leroy Cooke, for all their support over the past few years. In addition, he would like to acknowledge Thomas Victor Pickering, who is tragically no longer here with us in person but will remain with us in spirit. Yuan Xu thanks his wife Jing Song, son Anlan Xu, and daughter Antao Song for their unwavering and unconditional love. The research is partially funded by the National Key Research and Development Program of China (2022YFB3903700) and The Chinese University of Hong Kong (3136024).

Contents

List of Figures

Chapter 2

List of Tables

Chapter 1
Introduction

1. Global Climate Goals

Climate change is a significant threat to the world. Anthropogenic greenhouse gas emissions, mainly CO_2 from energy consumption, are its primary driver. Efforts to combat this threat culminated in the Paris Agreement, which was successfully negotiated in 2015 and entered into force in 2016. Within its framework, in the past few years, we have witnessed countries with more than 80% of global CO_2 emissions committed to achieving carbon neutrality in 2050, 2060, or 2070. If all net-zero goals were met, global temperature change could be limited to 1.8°C by the end of the century (Climate Action Tracker, 2021a). In total, 153 countries had set 2030 emission targets by 2023, and developed countries agreed to give US$100 billion of climate funding annually (UK GOV, 2021). Coal investment has decreased by 76% since the 2015 Paris Climate Accords (UK GOV, 2021). However, many countries, including China, India, Indonesia, and Australia, still incorporate coal into their development pipelines (Climate Action Tracker, 2021a).

Goals on methane reduction were also agreed upon, with 100 countries agreeing to a methane reduction pledge of 30% by

2030 (Climate Action Tracker, 2021a). Cooperation on land use and biological carbon sequestration was also mentioned, with 91% of world forests now being covered under a pledge to end deforestation by 2030 (UK GOV, 2021). In total, US$22.5 billion from government spending, donations, and private sector funding will be raised between 2021 and 2025 to support reforestation (UK GOV, 2021).

Within carbon markets, the EU carbon market saw carbon prices skyrocket in 2021, hitting over 84 euros per ton on December 16, 2021. Prices in the UK ETS market also hit a record high of nearly 80 pounds on the same day, although prices are volatile in both markets (Watson and Ram, 2021). The use of carbon markets is one of the critical ways the EU hopes to reach its 55% reduction in greenhouse gas emissions by 2030 and net-zero by 2050, with the cap on emissions decreasing by 2.2% annually between 2021 and 2030 (European Commission, 2021b). Updated prices are steady, with prices reaching around 90 euros per ton in 2023 (*EU carbon permits*, 2023).

Under the COVID-19 funding agreements, 600 billion euros is expected to go toward the EU's Green New Deal. This deal hopes to reduce European GHGs by 55% by 2030 compared to 2008 (European Commission, 2021a). President Biden has also reaffirmed net-zero in the United States by 2050 and, under the American Rescue Plan (2021), set aside US$30 billion for public transport funding and a further US$350 billion for state and local governments to reach their own climate goals (Climate Action Tracker, 2021b). Additional new NDC targets submitted just before the Glasgow summit align with Paris goals, if not the IPCC goals (Climate Action Tracker, 2021b). Additionally, in the United States, the Inflation Reduction Act of 2022 includes a subsidy of US$350 billion in investment for climate change goals.

Although this is promising, multiple countries have submitted 2030 climate targets that are weak and unlikely to be met, leading to doubts over whether their 2050 net-zero goals can be met. Countries such as Australia, Argentina, Brazil, India, Russia, and Saudi Arabia are among those unlikely to meet the climate goals necessary to limit climate change to an acceptable level (Climate Action Tracker, 2021a). Saudi Arabia's NDC states that renewables will supply 50% of electricity by 2030, but current generation stands at only 0.1%. Australia did not strengthen its 2030 climate goals at Glasgow and has already stated that it might exceed them (Climate Action Tracker, 2021a). As the world's fifth-largest carbon emitter, Russia has not pledged to cut coal or methane emissions, although, along with Brazil, it has pledged to end deforestation. This could be significant, as these two countries account for over 30% of the total forest area on Earth. Russia alone holds over 3.4 million km^2 of forest area, and Brazil has over 1 million square km^2 of forest cover (Duggal, 2021). This plan aligns with the Global Biodiversity Framework (GBF) signed in December 2022, which sets a target of protecting 30% of degraded ecosystems by 2030 (UNEP, 2022).

In 2014, China announced a goal to reach peak CO_2 emissions around 2030, which was significant in negotiating the 2015 Paris Climate Agreement. In September 2020, China further committed to carbon neutrality by 2060 at the UN 75th General Assembly. These goals, if met, are broadly in line with the International Energy Agency's (IEA) scenario projected to be compatible with the Paris climate goals (Varro and Fengquan, 2020). Under its new NDCs submitted in time for the Glasgow summit, China aims for carbon emissions to peak by 2030 and then achieve net zero before 2060 (NDC, 2021). Further, China plans to increase its wind and solar capacity to 1,200 GW by 2030, up from 414 GW installed in 2019, together with reforestation (NDC, 2021). Between 2015 and 2019, investments in renewables averaged around 490 billion yuan per year (NDC, 2021). Renewables (including hydro and nuclear) outnumbered new fossil-fuel installments four years in a row (2016–2019), making up 41.9% of China's power generation capacity (NDC, 2021).

Each province has made different climate goals under the 13th Five-Year Plan, ranging from 20.5% to 12% reduction of CO_2 intensity (NDC, 2021). Richer provinces such as Beijing, Tianjin, Shanghai, Shandong, and Guangdong have the highest emission control targets (NDC, 2021). Increasing thermal efficiency has been one method, with coal-fired power plants in 2019 having more than 20% drops in CO_2 emission intensity compared to 2005 (NDC, 2021). Improvements in heating appliances, investments in subway systems, and all-electric public transport in Shenzhen and Taiyuan were also registered by the end of 2019 (NDC, 2021). In July 2021, after many years of pilot programs, China launched the world's largest carbon market, with a market size of 4.5 billion tons of CO_2 (NDC, 2021).

Further, the United States and China signed a joint agreement at COP 26, creating a working group between the two countries to oversee emission cuts (The Guardian., 2021). However, China's current and expected emissions will likely still exceed the goals set for meeting the Paris Climate Agreement and the IPCC goals (Climate Action Tracker, 2021b). Further increased investment in fossil fuels was apparent in 2020, with 38.4 GW of new coal plants commissioned (Climate Action Tracker, 2021b).

2. Hong Kong and the Greater Bay Area: An Overview

This report focuses on the local energy infrastructures of Hong Kong, Macau, and Guangdong. However, each region imports significant energy and fuel quantities outside its boundaries from other provinces or overseas. Each location also has different climate goals. For example, in 2019, the Hong Kong government published a document on how Hong Kong could meet goals agreed to in the Paris Climate Agreement, which requires 80% of its electricity to be supplied by renewable or nuclear sources by 2050 (Chung, 2020).

Due to differences in size, transparency, and language, the information found for each area differs slightly. As Hong Kong and Macau are primarily service-based economies, these regions' energy and emissions have remained somewhat constant, with 92.2% of Hong Kong's GDP from the service sector in 2017 and 61.6% of Macau's GDP from the tourism and gaming sectors in 2016 (2022 Foundation, 2019). Guangdong's emissions drastically increased from 426 Mt CO_2 in 2000 to 610 Mt CO_2 by 2017 (Figure 1) (Zhou *et al.*, 2018a). Further, as Hong Kong and Macau are special administrative regions (SARs), publicly available data on how these areas produce, use, and store energy and fuel sources are more accessible than in Guangdong proper.

The Greater Bay Area (GBA) comprises nine municipalities in Guangdong Province, plus Hong Kong and Macau SARs. Guangdong has another 12 municipalities to the west, north, and east of the GBA. The GBA is one of the most vibrant economic engines in China and the world. Guangzhou and Shenzhen have caught up with Hong Kong regarding overall economic outputs (Figure 2). Although GDP per capita is still lagging, the gap has shrunk significantly. However, non-GBA Guangdong is quite different, with a much lower GDP per capita, much larger land and coastal areas, and a much lower density of population and economic activities. In 2020, the GBA had 64.3% of the total population (Figure 3) and 30.9% of the total land area (Figure 4), contributed 84.5% of GDP (Figure 2), and consumed 75.9% of electricity (Figure 5). According to the most recent data in 2019, GBA accounted for 58.2% of Scope 1 CO_2 emissions (Figure 1). The substantial regional disparity between GBA and non-GBA

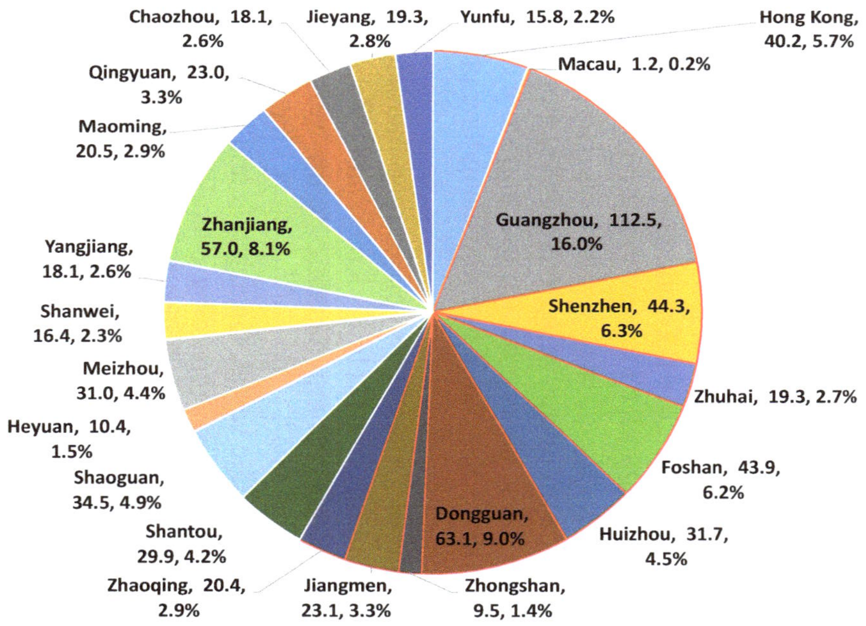

Figure 1. CO_2 emissions (unit: *million tons; Scope 1*) *by municipalities in 2020* (Shan *et al.*, 2022; Census and Statistics Department, 2023; Statistics and Census Service, 2023).

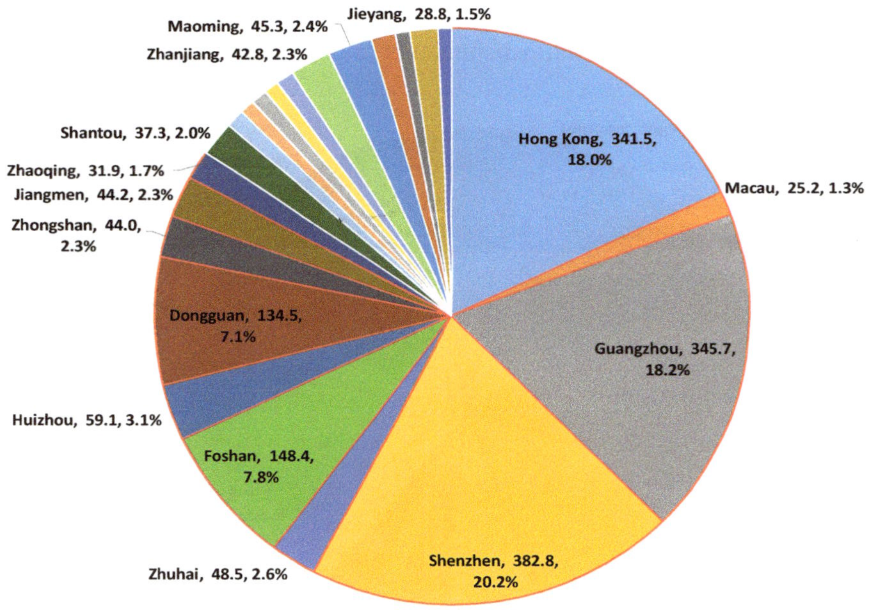

Figure 2. GDP in 2020 (unit: billion US dollars) by municipalities (red boundaries indicate those in the GBA) (Guangdong Bureau of Statistics and National Bureau of Statistics, 2022; Census and Statistics Department, 2023; Statistics and Census Service, 2023).

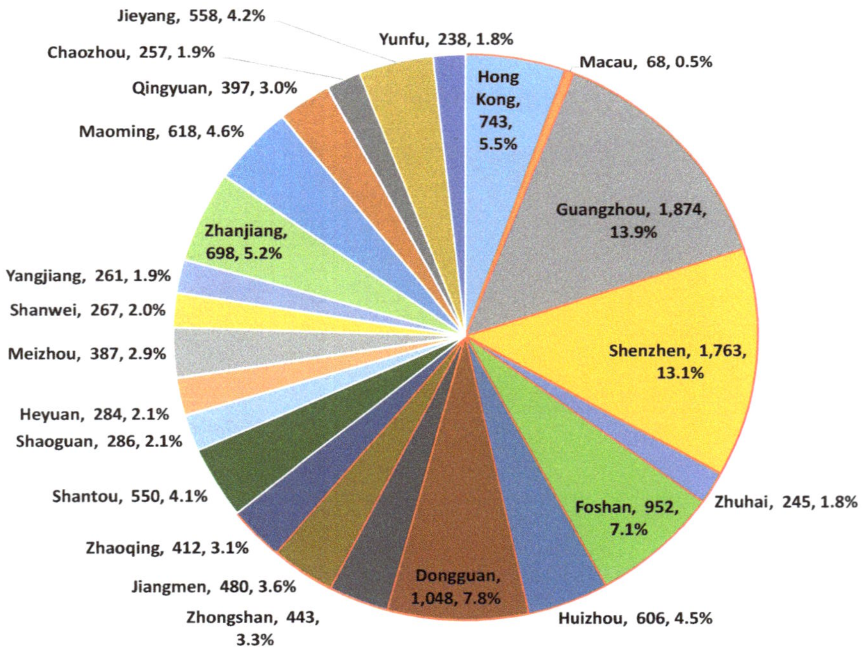

Figure 3. Population in 2020 (unit: 10,000 people) by municipalities (red boundaries indicate those in the GBA) (Guangdong Bureau of Statistics and National Bureau of Statistics, 2022; Census and Statistics Department, 2023; Statistics and Census Service, 2023).

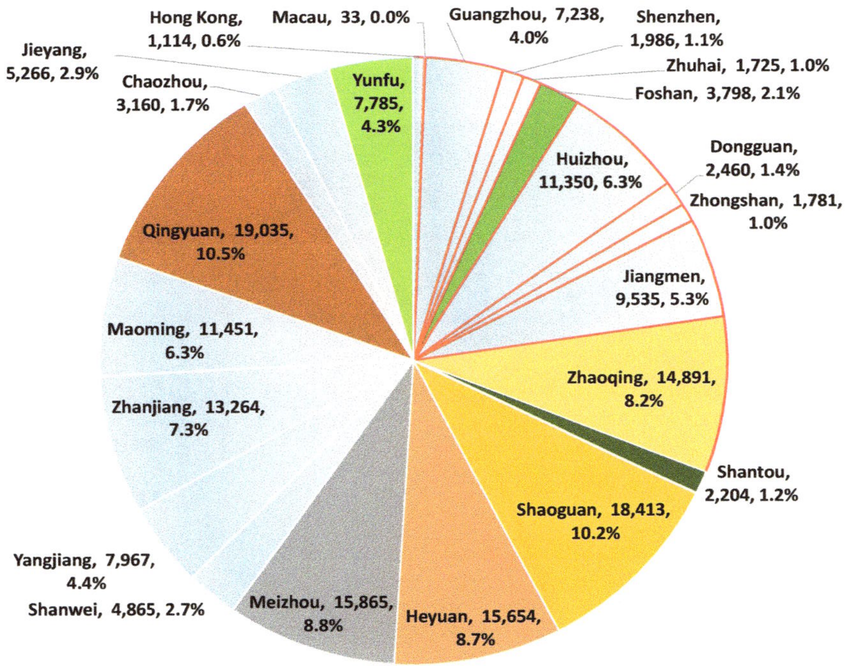

Figure 4. Land areas (unit: km^2) by municipalities (light-blue slices show coastal municipalities and thus with offshore wind potential) (Guangdong Bureau of Statistics and National Bureau of Statistics, 2022; Census and Statistics Department, 2023; Statistics and Census Service, 2023).

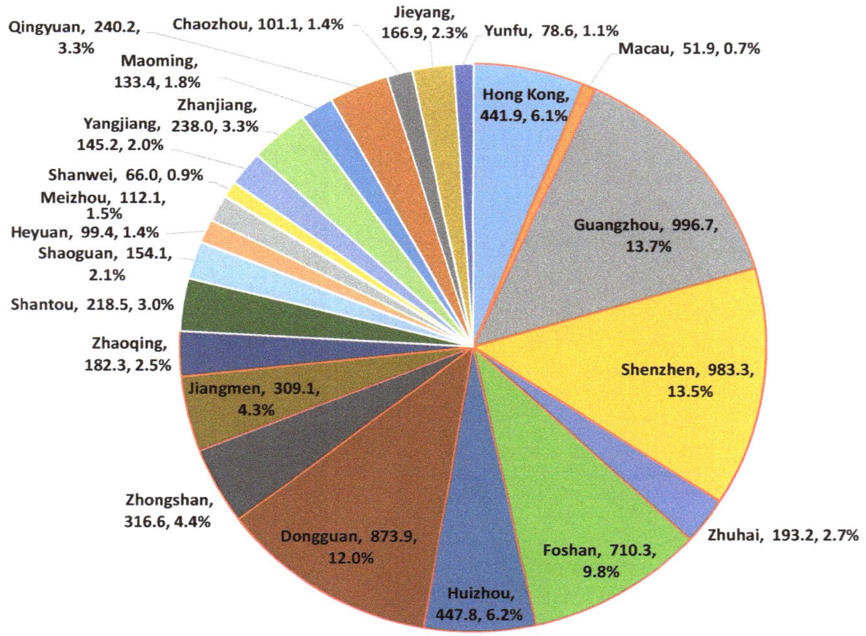

Figure 5. Electricity consumption (unit: TWh) in 2020 (Guangdong Bureau of Statistics and National Bureau of Statistics, 2022; Census and Statistics Department, 2023; Statistics and Census Service, 2023).

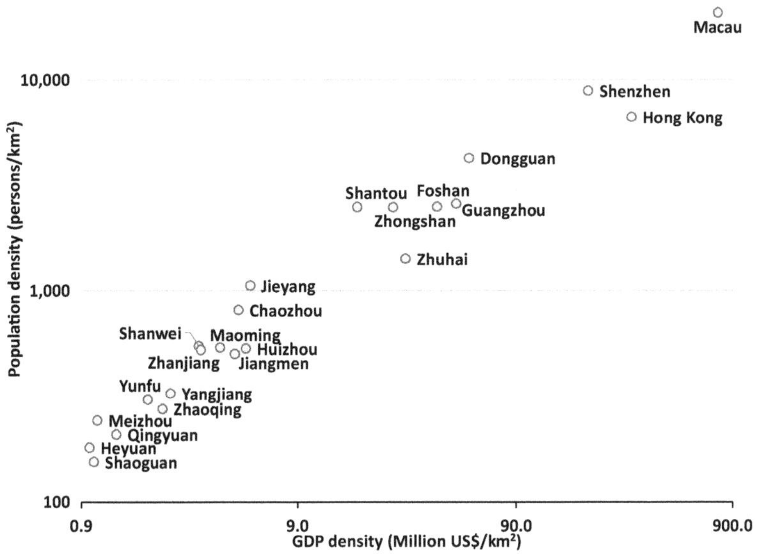

Figure 6. *GDP density and population density in 2020 for municipalities* (Guangdong Bureau of Statistics and National Bureau of Statistics, 2022; Census and Statistics Department, 2023; Statistics and Census Service, 2023).

Guangdong is similar to the contrast between eastern and western China. Carbon neutrality could become an excellent opportunity for narrowing this regional disparity, but it could also wide the disparity.

Using wind and solar energy requires substantial land to install wind turbines and solar PV panels. Although co-use is possible to a certain extent (such as on rooftops), much land should be primarily dedicated to their installation, especially under future carbon neutrality when easier opportunities are exhausted. As shown in Figure 6, the GBA tends to have a much denser GDP and population per km^2. On average, for solar PV at this region's latitude, 1 TWh of annual electricity generation requires about 7.5 km^2 of land (van de Ven *et al.*, 2021). Then, the annual

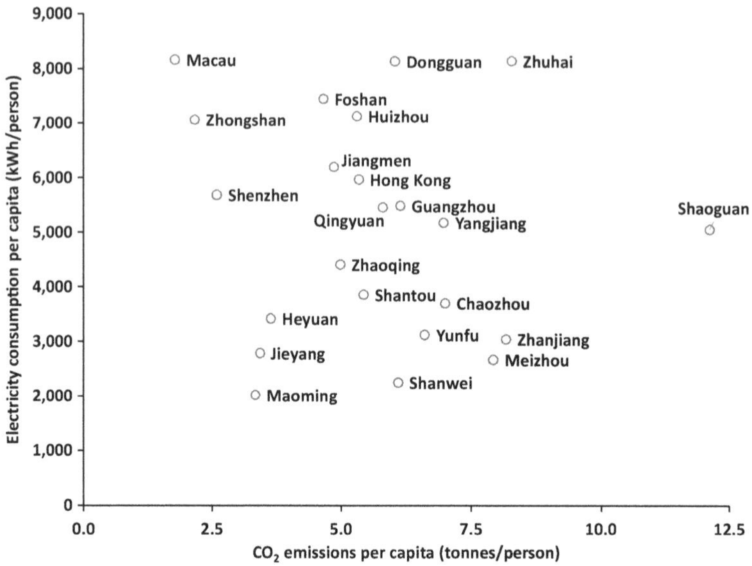

Figure 7. Electricity consumption per capita and CO_2 emissions per capita by municipalities in 2019 (Shan *et al.*, 2022; Census and Statistics Department, 2023; Statistics and Census Service, 2023).

revenue from solar electricity will be about a few million US dollars per km^2. At the same time, almost all GBA municipalities (except Zhaoqing, Jiangmen, and Huizhou) have much higher GDP densities than this threshold (Figure 6), indicating that, from an economic perspective, their land generally should not be used for dedicated solar PV installation. In contrast, most non-GBA municipalities, except for Shantou, produced much less economic output per km^2 than the expected annual revenue from solar electricity (Figure 7). The significant environmental impacts and land use of onshore wind in surrounding areas also suggest that they should be installed largely away from locations with high populations or GDP densities. Offshore wind also occupies significant coastal space, which may compete with sea routes, especially close to busy ports, while the GBA is home to three of the world's busiest container ports (Shenzhen, Guangzhou, and

Hong Kong). As a result, due to the high land or coastal space requirements, any significant solar or onshore and offshore wind developments within Guangdong will likely be outside the GBA proper.

3. Current Energy Systems

We collected detailed data on individual electricity generation units in Hong Kong, Macau, GBA Guangdong, and non-GBA Guangdong. The data for Hong Kong and Macau are complete due to their small sizes. For Guangdong Province, our data cover 13,970 MW of natural-gas-fired generation units and 54,336 MW of coal-fired units in 2015, while Guangdong's official energy statistics, which provided only aggregate total capacities, show that the numbers were 14,270 MW and 57,950 MW, respectively (Guangdong Provincial Government, 2022). For 2020, our data cover 25,122 MW of natural-gas-fired units and 65,746 MW of coal-fired units, compared to the official aggregate numbers of 28,380 MW and 64,270 MW (Guangdong Provincial Government, 2022). For all power plants, our data showed 84,641 MW in 2015 and 125,098 MW in 2020, whereas the official statistics were 98,170 MW and 141,770 MW, respectively (Guangdong Provincial Government, 2022). Accordingly, our detailed data form nearly a complete dataset of fossil-fuel-fired power plants and have only 13.8% and 11.8% discrepancies from official aggregate statistics in the two years, respectively. This high-quality dataset serves as a solid foundation for our analysis.

3.1. *Hong Kong*

When discussing energy infrastructure in Hong Kong, it is essential to note the nature of production and usage. Hong Kong is a net energy importer, having a primary energy requirement of over 598,000 TJ (terra joules) of energy in 2019 and a final energy requirement of 339,400 TJ of energy (HKGOV, 2020c). Like Mainland China, Hong Kong's energy requirement has not peaked, with 8.1% and 18.8% increases in Hong Kong's primary and final energy requirements, respectively, between 2009 and 2019, or annual average rates of change of 0.8% and 1.7%, respectively (HKGOV, 2020c). CO_2 emissions data on a per capita basis show Hong Kong's emissions tend to remain relatively constant, with per capita CO_2 emissions standing at 6.07 tons CO_2 in 1998, falling to 5.67 tons in 2001, peaking at 6.46 in 2005, and falling to 5.91 tons per person in 2017 (Ritchie and Roser, 2017). The Hong Kong government data show a similar trend, with total emissions in Hong Kong peaking in 2014 at 6.2 tons per capita and dropping to 5.4 tons in 2018 (HKGOV, 2020d). Installed capacity, peak demand, and electricity generated have remained nearly constant between 2009 and 2019 (HKGOV, 2020c). The maximum installed capacity in Hong Kong dropped slightly from 12,624 MW in 2009 to 12,225 MW in 2019. At the same time, peak demand fell from 10,153 MW in 2009 down to 9,601 MW in 2019. As such, total electricity generated per year is nearly constant, with generation at 178,888 TJ in 2009, peaking at 180,329 TJ in 2014, and falling to 177,033 TJ in 2019 (HKGOV, 2020c). Electricity generation accounts for around 65–70% of Hong Kong's greenhouse gas emissions, depending on the source (HKGOV, 2017a; Jiang *et al.*, 2020). Of the 12,225 MW installed, 8,988 MW is owned by Castle Peak Power Company Limited (of China Light & Power, or CLP) and the rest by Hong Kong Electric Company (HEC) Limited (CLP, 2020). In April 2019, CLP and HEC, two vertical monopolies with distinct service areas in Hong Kong, signed two schemes of control agreements (SCAs), reflecting the governments' commitments to fighting climate change and meeting public aspirations until 2033 (HKGOV, 2019).

The SCAs highlight and affirm each power company's role in their share of power generation, ensuring energy security and reasonable energy prices in Hong Kong for the next decade (HKGOV, 2019). Regarding emission intensity during 2014–2019, CLP's total carbon intensity per kWh decreased from 0.64 kg CO_2 to 0.50 kg CO_2, with a goal of 0.4 kg CO_2 for 2020 when new gas-fired units come online (CLP, 2020). CLP's Hong Kong branch employs 4,689 people in part-time or full-time roles as of year-end 2020 (CLP, 2021). Of these, 3,910 are engaged directly in Hong Kong's regulated electricity business, 417 in Hong Kong's non-regulated electricity business, and 362 through CLP holding companies also based in Hong Kong (CLP, 2021). Hong Kong Electric employs 1,713 permanent employees as of 2020 (HKEI, 2021). Hong Kong Electric does not state how many part-time employees it has. Therefore, a conservative estimate is that 6,040 people are employed via the power industry within Hong Kong and 362 more through power investment companies via CLP holding company. Hong Kong's total employment at year-end 2020 was 3.65 million, with a labor force of 3.88 million. Both had decreased due to COVID-19 by 5.1% and 2.2%, respectively (HKGOV, 2021b).

Therefore, the power industry represented 0.17% of Hong Kong's employment. However, as taxable revenue, the two power companies together delivered nearly HK$2.4 billion in tax to the Hong Kong government in 2020, with 541 million from HEC and 1,855 million from CLP (CLP, 2021; HKEI, 2021). The heating industry run by the company, Towngas, is explored in much greater detail in Chapter 2; however, it needs mentioning here as well since the primary gas used in heating within Hong Kong is town gas, which is 61% natural gas and one of the primary energy fuels used in Hong Kong. The company employed 2,130 people in the gas business at the end of 2020 and 2,495 people within Hong Kong and paid HK$1.73 billion in tax in 2020, down from HK$2.29 billion in 2019 (Towngas, 2021). In total, the energy and heating industry employed just under 8,900 people, or 0.24% of Hong Kong's total employment in 2020, with

HK$4.13 billion given in tax in 2020, although this was down significantly from the 2019 tax of HK$5.69 billion due to decreased profits (CLP, 2021; HKEI, 2021; Towngas, 2021). Therefore, the heating and energy industry has an oversized impact on Hong Kong's taxable revenue. In 2019–2020, Hong Kong's corporations paid HK$155.9 billion in tax, while the total revenue collected was HK$303.6 billion (down 11.1% from the previous year). Although the corporations only employed 0.24% of Hong Kong's workforce, their tax revenue to the government was 2.65% of corporations' taxes received, or 1.36% of all taxes received in 2019–2020 (HKGOV, 2020a). This was down significantly even from the 2019 amount of 3.42% of all corporation tax received and 1.67% of all tax received (HKGOV, 2020a).

CLP recently published its new carbon plans, calling for a final intensity target of 0.15 kg CO_2/kWh by 2050 (CLP, 2019). HEC had a higher carbon intensity than CLP due to the lack of nuclear power (Shenzhen Daya Bay Nuclear Plant) in the former's energy mix. In 2018, its carbon intensity was 0.80 kg CO_2/kWh, with the projected intensity decreasing to 0.60 kg CO_2/kWh by 2023 with the introduction of three new gas-fired units (Hong Kong Electric, 2022).

Although showing slightly different numbers, other sources agree with the same trend, with Hong Kong's per capita emissions falling from 6.46 tons CO_2 per capita in 2007 to 6.10 tons CO_2 in 2018 (Knoema, 2017). However, this emissions data only account for Scope 1 and 2 emissions (Scope 1 emissions are released within the city, while Scope 2 emissions include emissions released from imported power into the city). If Scope 3 (consumption-based) emissions are added, expected emissions will increase by about a factor of three. Hong Kong's Scope 3 emissions could be as high as 34.6 ± 6.3 tons of CO_2 per capita, while its total emissions of 208.5 ± 37.8 million (Mt) tons of CO_2 make the city the fourth highest in the world and the highest per capita globally (Moran *et al.*, 2018). This high level of emissions has made Hong Kong issue multiple climate plans with different focuses and timescales, specifically the Energy Saving Plan (2015–2025), the Hong Kong Climate Action Plan 2030, and the Clean Air Plan 2013–2017 (HKGOV, 2015, 2017). Government data for 2017 and 2018 show 40.4 Mt and 40.6 Mt of CO_2 emissions, respectively, with both years seeing 26.6 Mt CO_2 from the power sector and 7.36 (2017) and 7.35 (2018) Mt CO_2 from the transport sector (HKGOV, 2020b).

Regarding CO_2 emissions, the energy-saving program does not mention Scope 3 emissions, while the Climate Action Plan alludes to the need to address consumption-based emissions (HKGOV, 2017b). However, the current policy scenario still sees emissions of 22 Mt CO_2 by 2050, while groups such as the World Resource Council have suggested much more ambitious decarbonization plans to bring that amount down to 3.9 Mt CO_2 (Jiang *et al.*, 2020). This decarbonization plan would require an annual change of 6.6% from 2018 to 2050 (Jiang *et al.*, 2020). For climate budgets, the Hong Kong government has spent HK$47 billion on different projects over the last 10 years to counter climate change and expects to spend a further HK$240 billion over the next 15–20 years (HKGOV, 2021a).

3.2. *Macau*

Unlike Hong Kong, Macau receives most of its electricity from outside. Companhai de Electriciade (CEM), Macau's primary power producer, showed that, in 2019, Macau imported 86.3% of its energy from the China Southern Power Grid (CSG). CEM generated a further 10.3%; the rest was purchased from the Macau Refuse and Incineration Plant (CEM, 2020a). Macau would be unable to meet its energy demand as its operation capacity has been 408 MW since 2016; however, peak demand from the same time has ranged from 955 MW to 1,062 MW

(CEM, 2021). About 72% of its energy usage was in the commercial sector in 2020 and 25.7% in the residential sector (CEM, 2021). Employees at CEM numbered 710 in 2020 and have remained relatively steady since 2011, only differing by 43 people in 10 years, with the majority being full-time staff (CEM, 2021). Tax income was just under 64 million MOP (about HK$62 million) for 2020. Overall, CO_2 emissions per person are estimated at around 3.37 tons (Ritchie and Roser, 2017). Macau's GHG emissions peaked in 2005, although emissions from the transport sector increased (EPPM, 2012). Between 2000 and 2019, electricity consumption increased from 1.57 TWh (terawatt hours) to 5.81 TWh. Imported electricity amounted to 4.98 TWh, with Macau's energy reliance increasing over time since. In 2004, Macau only imported 1.51 TWh but used 1.9 TWh of energy (Government, 2020). Total direct and indirect emissions through power generation in Macau for 2019 were 4.69 million tons, of which 4.21 million tons were from indirect emissions produced via CSG's power grid (CEM, 2020b).

3.3. *Guangdong*

Guangdong's population has more than doubled over the previous 40 years, from 52.3 million in 1980 to 111.7 million in 2017 (2022 Foundation, 2019) – a 113% rise in population, well above China's simultaneous 41.3% rise (2022 Foundation, 2019). Hong Kong and Macau import significant amounts of power from Guangdong, with Guangdong providing most of Macau's energy (Zhou *et al.*, 2018c). In 2016, Guangdong was estimated to produce 610 Mt of carbon emissions, up from 426 Mt in 2000 (Zhou *et al.*, 2018c). Emission intensity per unit of GDP (in this case, 10,000 yuan) ranges from 0.10 tons of CO_2 for Macau up to 2.9 tons for Meizhou (Zhou *et al.*, 2018c). Four of the 23 municipalities as of 2015 had begun to decrease emissions, 11 had peaked, and the rest still had rising emissions (Zhou *et al.*, 2018c). The economic outputs of Guangdong's electric power, gas, and water supply reached 224.9 billion RMB in 2019 (Guangdong Statistics, 2020).

Guangdong Energy Group operates coal, hydro, natural gas, wind, and solar power plants throughout Guangdong. It is a state-owned corporation with 33.1 GW of installed capacity, primarily in Guangdong, making it the largest energy group in the province. The group controls 21 thermal power plants (29.4 GW), 13 onshore and 7 offshore wind farms, 10 solar farms, and 11 hydroelectric power plants as of 2019 (Guangdong Energy Group, 2019). It also controls the Guangdong Electric Power Development Company. The group has over 20 GW of coal-fired plants in Guangdong, owning 35 power plants, of which 15 are 1 GW or above. Guangdong Energy is becoming a prominent player in the natural gas market, with 6 GW of LNG plants and a further 2 GW under construction in Guangdong (Guangdong Energy Group, 2019). The company operates 636 km of LNG pipeline in Guangdong, with an additional 500 km under construction. The group is also one of the more prominent players in Guangdong's renewable energy sector, with a 100 MW biomass powerplant, 734 MW of solar (distributed), 632 MW of onshore wind, and the construction of seven offshore wind farms in Guangdong at 2.2 GW, or 44.5% of Guangdong's total as of 2019 (Guangdong Energy Group, 2019). At the end of 2019, the group had a total asset value of nearly 146 billion RMB and just under 14,500 employees (Guangdong Energy Group, 2020). However, other companies in Guangdong operate interprovincially, so the division of assets and employment is hard to determine. China Southern Power Grid (CSG) handles electricity transmission for Guangdong, Guangxi, Hainan, Yunnan, and Guizhou (CSG, 2019). According to Fortune, employment within the company was around 289,000 as of the middle of 2021, with an asset value of US$155.2 billion. Fortune ranks the company at 91 out of the Global 500, up from 316 in 2005, the first year on record (Fortune, 2021).

Chapter 2

Fossil Fuel Infrastructures

1. Global Trends

Coal poses a particular challenge to emission reduction goals, with predictions in late 2020 suggesting that up to 60% of the current coal-fired power generation fleet could still operate in 2050 (IEA, 2020c). Coal holds both long-term sunk costs and short-term geopolitical concerns around the price of alternative fuel types, as well as energy security concerns. Within China and Southeast Asia, the average age of the plants is under 13 years and under 20 years, respectively, meaning that the return on investment (ROI) if the plants were retired would be low (IEA, 2020c). Although China was a major international funder of coal power plants under the Belt & Road Initiative, it announced in late 2021 that it would discontinue this practice; moreover, it did not fund any new overseas coal-fired plants in the first half of 2021 (Gunia, 2021). Geopolitical events can also increase coal-fired plants' lifespan in the short term, especially in certain regions. The Ukraine–Russia war led to a significant spike in energy pricing throughout Western Europe, especially allowing for the reclassification of natural gas as a green fuel and the recommission and extension of coal-fired power plants in Germany (Wrede, 2022). This geopolitical situation, however, is seen as a more short-term concern, with the lifting and recommissioning of coal plants expected to last only until March 2024 (Wrede, 2022).

Globally, coal demand has risen slightly between 2010 and 2021, from 5.22 billion tons to 5.64 billion tons (IEA, 2022). Usage is focused on power production (64.5%) and industry (28.8%) (IEA, 2022). Under all net zero goals, coal usage should fall. However, the reality differs significantly, with the global net zero by 2050 policy seeing only 3.02 billion tons of coal used by 2030 (IEA, 2022). Regionally, North America and Europe have seen the most significant decrease in coal from 2010 to 2021 (768 to 389 Mt and 539 to 369 Mt, respectively) (IEA, 2022). Coal usage in Central America, South America, and Russia all increased slightly. At the same time, coal usage in the Middle East was low, at 5 million tons in 2010 and 2021. Coal's global rise is primarily from the Asia-Pacific region, which increased coal usage by over 800 million tons annually between 2010 and 2021, an average rise of over 72 million tons per year (IEA, 2022).

Oil demand, especially for aviation and shipping, decreased significantly over COVID-19 (IEA, 2022). In 2021, daily oil consumption in these two industries was 20% lower than before COVID-19 (IEA, 2022). Meanwhile, uncertainty regarding demand and sanctions on Russia still leads to prediction difficulties (IEA, 2022). Total production and supply, especially in 2021, were low, with the number of new oil resources being discovered in 2021 at its lowest since the 1930s, while refining capacity fell for the first time in 30 years (IEA, 2022). The IEA sets three primary predictions for total oil demand in 2030. The first stated policy is for consumption to meet 102 million barrels per day (mb/l) by 2030 (IEA, 2022). While under announced pledges, demand will peak soon and then decrease to 92 million bbl/day, a similar level to where we were in 2021 (94.5 million bbl/day) (IEA, 2022). However, both are still far from the demand required under the global net zero 2050 goal, which would require oil demand to drop to only 75 million bbl/day in 2030 (IEA, 2022).

Natural gas demand has slowed, with demand under the current stated policies seeing a 5% increase in gas demand from 2021 to 2030 (IEA, 2022). This is compared to a 20% rise from 2011 to 2020 (IEA, 2022). Under the announced pledges, however, demand is expected to be 10% lower in 2030 compared to 2021, and under net zero goals, demand will be 20% lower (IEA, 2022). Gas demand is expected to decrease in the developed world and increase in the developing world (IEA, 2022). Regions such as the EU have seen drastic changes in gas supply, including the Russian gas exports to the area during 2021–2022, helping to drive infrastructure changes with an expected fall in demand of 40% by 2030 (IEA, 2022). Emerging Asian markets are expected to see a 20% increase in demand, with most new demand met by LNG (IEA, 2022).

2. Coal Infrastructure

2.1. *Coal consumption*

China produced nearly half of the global coal in 2019 while importing 300 million tons in the same year (Our World in Data, 2020; Mullen, 2021). Total production reached 3.84 billion tons in 2020 (Mullen, 2021). Coal production in China initially peaked in 2013, falling by 10.7% by 2016, but by 2019, it had risen back to its old 2013 peak level (Our World in Data, 2020). In 2021, we saw coal demand at 3.16 billion tons (IEA, 2022). Coal will likely remain an essential part of the Chinese energy market for the foreseeable future, with the China National Coal Association proposing limits on coal usage at 4.2 billion tons per year by 2025; moreover, during COP26, China, along with India, did not sign the phase-out plan for coal usage (Mullen, 2021;

Simon, 2021). However, as a share of energy consumption, coal has decreased from 70.2% of the total energy used in 2010 to 57.6% in 2019 (Our World in Data, 2020).

Coal in Hong Kong is almost exclusively used to generate electricity. The Hong Kong government has since 1997 tried to decrease the amount of coal in the energy mix, first by banning the construction of any new coal-firing power stations in 1997 and by slowly phasing out Hong Kong's reliance on coal (HKGOV, 2017c).

In 2019, the consumption of coal in Guangdong was estimated at 168.34 million tons, divided chiefly between 55.08 million tons (32.7%) used in the manufacturing industry and 111.59 million tons in the power and heating industry (66.3%) (Guangdong Bureau of Statistics and National Bureau of Statistics, 2020). This is slightly higher than the country's ratio, with about 60% of its coal in 2018 used in the power sector (EIA, 2020). Coal's composition in the Guangdong energy mix has decreased over the past decade, from 50.2% of the primary energy requirement in 2010 to 34.2% of the primary energy requirement in 2019 (Guangdong Bureau of Statistics and National Bureau of Statistics, 2020), but there has been a rise in total energy consumption from coal. Only 5.6 GW of coal plants were retired in Guangdong between 2000 and 2020, while 56 GW of coal plants were constructed (Global Energy Monitor, 2020b). This change in usage mirrors China's, which saw an increase in coal consumption due to increased demand from energy markets, with China accounting for about half of global coal demand. While between 2000 and 2020, China retired 102.8 GW of coal plants, as of July 2020, the construction of over 150 GW of coal plants is either underway or has received permission (Global Energy Monitor, 2020b). The slight decrease in coal usage is lower than that called for by the 13th Five-Year Plan, which initially wanted to decrease coal usage from 64% of the primary energy requirement in 2015 to 58% by 2020 (National People's Congress, 2016). As of July 2020, Guangdong operated over 74.1 GW of coal-fired power capacity, while China in total had over 1,000 GW of coal power, with a further 98.5 GW under construction (Global Energy Monitor, 2020b).

2.2. *Coal transport infrastructures*

Coal prices in Guangdong between 2014 and 2019 averaged 10–20% higher than those in China. Guangdong removed all coal mining capacity in 2006 while importing roughly 180 million tons annually (Liu and Jin, 2020). The import ratio is about 30% from abroad (50 million tons, or 16.7% of China's total imports) and 70% from China's north and northwest regions (Liu and Jin, 2020). Foreign imports into Guangdong had been mainly from ASEAN (55.8%), Australia (26.5%), and Russia (14.5%) (Wang and Ducruet, 2014). Import capacity is also high, with Guangzhou port alone being able to handle 60 million tons per year in 2017, more than the amount currently imported from foreign sources for the entire province (Reuters, 2017). The Pearl River Delta, in total, had an unloading capacity of 75 million tons in 2012 (Wang and Ducruet, 2014). Proportionally, Guangdong imports a high amount of foreign coal compared to the rest of the country (Xu and Maguire, 2020). At the same time, preliminary results showed that coal trade dropped 10–12% globally (10% for thermal coal and 12% for metallurgical coal) in 2020 due to COVID-19 (IEA, 2020b). The international supply chains was disrupted by a slowdown of Australian coal imports in late 2020, causing blackouts throughout Guangdong (Mudie, 2020). Although, in this case, the slowdown in imports was driven by political restrictions, it does highlight the concern for energy security (Mudie, 2020).

Regarding the transport of coal within China, traditionally, rail has been an essential vector, accounting for 60% of China's coal transport, reaching over 2.3 billion tons in 2013 (Cui *et al.*, 2018). Coal traffic by rail increased by 145.8% from 2000 to 2012 (Cui *et al.*, 2018). Coal transported through coastal ports in China was 1,163 Mt, and inland waterways shipped 484 Mt in 2010 (Jianjun, 2013). Rioux *et al.* (2016) estimate that transport by road was 20% of output (920 Mt) and that by inland waterway was 184 Mt (5%), coastal ports handled 620 Mt, and most of the rest was by rail. Data from 2020 also suggest a significant surge in rail transport, with 5.6 million tons of coal daily (173.6 million tons total) being moved in December 2020, 10% higher than in December 2019 (State Council, 2021). Transport bottlenecks are also heavily linked to a lack of rail infrastructure or environmental conditions that slow down transportation (Jianjun, 2013; Rioux *et al.*, 2016; Cui *et al.*, 2018), which caused price increases in coastal provinces, costing the Chinese economy 105 billion RMB in 2013 (Rioux *et al.*, 2016). For example, extreme winter weather conditions in 2010–2011 caused bottlenecks in the production and transportation of coal from northern provinces to southern provinces, leading to an increase in overseas imports (Cornot-gandolphe, 2014).

2.3. *Coal-fired power plants*

Hong Kong currently operates two large coal-fired plants: one on Lamma Island (Lamma Power Station by HK Electric) and the other in Tuen Mun (Castle Peak by CLP). Both plants were built in the 1980s and produce power through a mix of gas and coal, with a gradual shift toward gas. Castle Peak has a coal generation capacity of 4.1 GW, while Lamma uses a mix of natural gas and coal with a rated coal capacity of 2.25 GW (HK Electric, 2014). Lamma's coal capacity has recently dropped, as HK Electric increased the ratio of gas-powered generation at Lamma from 32% in 2018 to 70% by 2023 (Electric, 2019). Two coal-fired units were retired in 2017 and 2018.

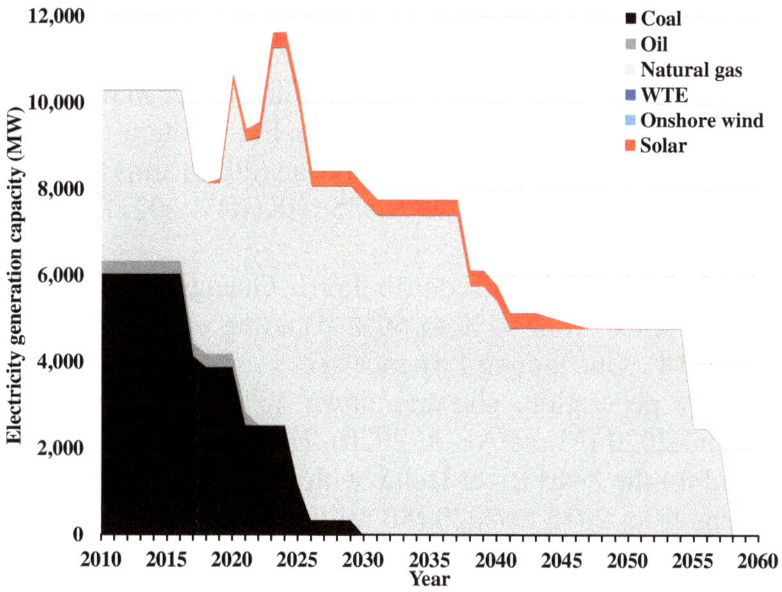

Figure 1. Electricity generation capacity in Hong Kong (numbers after 2022 include those planned and retiring).

Coal accounted for 48% of Hong Kong's electricity supply in 2015, with government plans aiming to reduce this to 25% by 2020 (Figure 1) (HKGOV, 2017c). Data show that coal usage had been reduced from 2000 to 2009 before remaining relatively steady for the next few years (Jiang *et al.*, 2020). Import data from 2014 onward show a slight drop in coal usage, with Hong Kong importing 13.79 million tons of coal in 2014, dropping to 10.04 million tons by 2019 (HKGOV, 2020). Coal's share in primary energy requirement in 2014 was 56.7% and dropped to 43.1% in 2019 (HKGOV, 2020). In total, 97.6% of the coal imported into Hong Kong was used in electricity generation, with the rest kept as stock (HKGOV, 2020). In 2019, 78.6% of coal in Hong Kong was from Indonesia, and 11.3% was imported from Russia. However, Hong Kong's coal-fired power plants are relatively inefficient, both running as subcritical plants due to their advanced age; Castle Peak, at 37 years old, emits about 22.3 Mt of CO_2 per year, and Lamma emits 11.4 Mt of CO_2 per year. The emission factor could be over 50% higher per MWh than the ultra-supercritical plants in Guangdong. For medium-term goals, the new Hong Kong Climate Action Plan (2021) aims for no coal use in electricity generation by 2035 (HKGOV, 2021).

During 2000–2017, coal's contribution to Guangdong's CO_2 emissions increased from 47% to 60% (Daiqing *et al.*, 2013; Zhou *et al.*, 2018). Guangdong Province seeks to reduce its coal-fired electricity generation, shutting down 13 coal plants between 2018 and 2020 (Mirae Asset, 2020). Higher reduction rates are planned for the Pearl River Delta, with a decrease of over 10 million tons from 2015 to 2020 (80 million to 69.7 million) (Mirae Asset, 2020). As of July 2020, 52 coal-fired plants are present (including all types in construction or operating) in Guangdong, with a total capacity of 74.1 GW, an average plant size of 1.43 GW, and an average age of 13 years (Global Energy Monitor, 2020a).

Space requirements: Coal-fired power plants require space for conducting operations. Space is necessary for the generators, storage units, and wharfs if the plant is primarily fed by ships, among other considerations. An analysis of the space required using satellite images of the Hengyun C+D coal plant, the Shajiao A+C power plant, Castle Peak power plant, Goushan and the Haifeng power plants was completed to gauge the area requirements for these power plants. Each of these power plants produces the majority of its power output from coal usage, is located along the coast, and is fed by ships. Area measurements encompassed both the onsite land area taken up by the plant and the area of the offshore wharf, with a buffer of around 15 m on either side of the pier as an assumption that the sea area is a restricted zone. For larger plants, the breakwater line was not included in the calculations.

The subcritical Hengyun C+D coal power plant took up about 685,000 m^2 with a rated capacity of 1,080 MW, which leads to 634 m^2/MW of capacity. This is compared to the space requirements of the ultra-supercritical Haifeng plant of 2,100 MW, where its requirement of 727,000 m^2 leads to a ratio of 346 m^2 per MW capacity. The Castle Peak plant in Hong Kong holds an area of 1.05 km^2 with a rated capacity of 4,110 MW, leading to a 255 m^2/MW ratio. The other power plant in Hong Kong, the Lamma Power Station, holds an area of 1.76 km^2 with a rated capacity of 4,280 MW coal- and gas-fired units, for a ratio of 421 m^2/MW. As the most significant coal-fired power station in Guangdong, the seven-unit Goushan plant has a total area of 1.96 km^2 with a rated capacity of 5,000 MW, leading to a 392 m^2/MW ratio. However, it is also important to note that as plants age, specific units are retired in succession, which can increase the ratio of area to capacity as the units decommissioned may not be conveniently converted into alternative usage. Castle Peak is an example, as the plant is subdivided into Castle Peak A and B, with Castle Peak A expected to be decommissioned between 2022 and 2025 and B before 2030 (CLP, 2021). Lamma Island is a similar case in which Hong Kong Electric is slowly decommissioning its coal-fired units and replacing them with gas-fired units, increasing the amount of gas used in the plant from 30% in 2017 to 50% in 2020 (Hong Kong Electric, 2022).

Capacity factors: Due to a significant increase in renewables between 2015 and 2020, coal power plants have been underutilized in China as grids prioritize cleaner electricity sources (Xu and Stanway, 2021). In China, coal plants, on average, run for less than 4,000 h a year (10.95 h a day) compared to their designed level of 5,500 h (15.06 h a day) and much less than the theoretical maximum (Xu and Stanway, 2021). Other data support this, with load factors in China's fleet dropping from over 60% in 2005 to 46% in 2016 (Spencer *et al.*, 2017). Low load factors also heavily influence revenue because of lost income when a plant is not operating and the high operating costs associated with restarting a plant (Sugden-Nalbandia, 2016). Carbon Tracker's study (Gray, 2020) suggests that most coal plants would not be profitable at these levels. In the United States, for example, in 2008, the expected restarting cost for a typical 500 MW coal plant would range from US$93,600 for a hot start (offline 1–23 h) to over US$173,900 for a cold start (offline >120 h) (Sugden-Nalbandia, 2016). However, these costs are based on data from the United States, and cost per MW decreases as coal plant generator size increases; this still represents a significant issue in profit generation (Sugden-Nalbandia, 2016). For Guangdong, even before the substantial increase in renewables, coal utilization was below the national average of 4,028 h, averaging only a little more than 2,000 h (Sugden-Nalbandia, 2016). Even though emissions have increased, the average intensity of emissions per kWh for coal has decreased. Between 2000 and 2020, the carbon intensity of coal plants in China has reduced from over 1,100 g of CO_2 per kWh to about 900 g of CO_2 per kWh by 2018 (IEA, 2020a). In part, this is because the thermal efficiency of China's coal fleets increased from 30% in 2000 to 39% by 2018 (IEA, 2020a). Overall, Guangdong's thermal (coal, gas, and oil) power generation efficiency has increased from 31.13% in 1990 to 41.58% in 2019 (Guangdong Bureau of Statistics and National Bureau of Statistics, 2020). At the same time, new ultra-supercritical plants can have efficiencies as high as 48% (Zhang *et al.*, 2017). The benchmark value for a 1,000 MW plant in Guangdong for 2017 and 2018 was 800 g of CO_2 per kWh (Liu and Jin, 2020). Even so, on average, CO_2 emissions from a 500 MW coal-fired plant are 2.37 times greater than those for a comparable gas turbine (Zhou *et al.*, 2018).

Regulation: Nationally, about half of the coal plants are loss-making, with utilization at 49% in 2019 due to the relaxing of rules in 2014, fueling a boom in new builds as provinces tried to boost short-term GDP growth through construction projects.

Regulation loosening to increase energy security is apparent, even though the regulation was retightened, with construction restarting on 18 GW of new builds and 37 GW of inactive projects revived in 2019 (Myllyvirta *et al.*, 2020). Both China State Grid and China Electric Power stated in 2019 that provinces might need to maintain or increase coal capacity to avoid shortages (Myllyvirta *et al.*, 2020). The most significant regulation hurdle, however, will be if China retains or tightens the current cap of coal power generation (1,100 GW) in the 14th Five-Year Plan (2021–2025) (Baiyu, 2021). Expert opinion is that to meet the 2030 goals, China would need to decrease capacity to 680 GW to be in line with the IPCC's 1.5°C goal or even peak emissions, while even meeting 1.8°C or 2°C goals would also require a slight reduction in capacity (Myllyvirta *et al.*, 2020; Baiyu, 2021). Regulations placed in 2019 do emphasize the need for coal to reduce emissions. They ban investment (domestic and foreign) in wet-cooled coal power generators with coal consumption greater than 300 g of coal equivalent per kWh (or the thermal efficiency should not be less than 41.0%) and air-cooled generators greater than 305 g of coal equivalent per kWh (Zhang *et al.*, 2020). Investment is prohibited in new coal plant construction that does not meet national standards (Zhang *et al.*, 2020). Further, new captive coal plants (such as power plants for steel mills; they can operate off-grid) are banned in the Pearl River Delta region (Zhang *et al.*, 2020).

Efficiency and cost: Fuel cost is a significant expense for coal plants; a gradual phasing out of less efficient plants is necessary for pollution control and managing operating expenses. Efficiency is critical in this respect, with China's 1,000 MW ultra-supercritical Guodian Taizhou II unit reaching an efficiency rate of 47.82%, the highest in the world for double-heated units (Power, 2017). This is impressive because of the thermal power generation process and thermodynamics laws; 60% is the highest possible efficiency level (Wang *et al.*, 2021).

This increase in efficiency has led to a decrease in coal equivalent consumed per kWh of electricity from 370 g in 2005 to 318 g by 2014, while power plants with a 43.5% efficiency can have coal usage amounts as low as 282.5 g per kWh (Zhang *et al.*, 2017; Wang *et al.*, 2021).

The increase in efficiency in Chinese coal plants can be seen when comparing the 100 most effective coal plants in China versus the United States, with the top 100 units in China using 286 g of coal equivalent per kWh, while in the United States, this is 375 g per kWh (Hart *et al.*, 2017). The highlight here, though, is the difference in average intensity between subcritical, supercritical, and ultra-supercritical plants (Hart *et al.*, 2017). Although not accounting for the number of hours used per plant, Carbon Brief (2020) measured the estimated emissions from 41 of the 50 power plants in Guangdong; of those counted, 24 were subcritical, 7 were supercritical, and 10 were ultra-supercritical. Guangdong's supercritical plants produced 17.8% fewer emissions than subcritical plants. In comparison, ultra-supercritical plants had 39.2% fewer emissions than subcritical plants (Carbon Brief, 2020).

As such, the upgrading of coal plants, especially with new technology to make them more efficient, can be a sound investment. Optimizing even efficient (supercritical) generators can reduce coal consumption and decrease the cost of electricity generation with minimal capital cost (Wang *et al.*, 2021). Wang's study showed that for a 600 MW supercritical unit, an additional US$0.94 million investment over a conventional system could increase net income by US$4.95 million from increased load and efficiency and reduced coal consumption (Wang *et al.*, 2021). Retrofitting a 320 MW subcritical unit at Xuzhou increased efficiency to 43.56% from 38.6% while allowing usage at load levels as low as 19% for more flexible generation (Patel, 2020). Even for larger 600 MW units, the total cost of the retrofit was

US$50 million, which is about a third of the cost to build a new unit (Patel, 2020). Therefore, this will be the most likely way of extending the lifetime of these middle-aged subcritical plants to avoid extremely high stranded asset costs.

Stranded assets: The increase in ultra-supercritical coal-fired power plants can help achieve short-term carbon intensity reduction. However, if Guangdong and China must begin phasing out coal use to approach carbon neutrality, the recent increase in coal generation capacity will become an ill-advised investment decision (Zhang *et al.*, 2017). It is likely that newer coal plants will not meet acceptable rates of return (Spencer *et al.*, 2017). Additionally, the IEA has also said there is no economic sense in building more coal plants, with bankruptcies already occurring within the sector (Carbon Brief, 2020). China has already mandated strict coal usage guidelines and emissions for new plants, allowing the building of only super- and ultra-supercritical plants (Hart *et al.*, 2018). This report believes Guangdong should decommission all subcritical plants within five years and halt new plant proposals. If no new coal plants were built past 2015 and assuming 30-year lifespans, the stranded assets for all of China were estimated to be between US$64.2 billion and US$97.8 billion (Spencer *et al.*, 2017).

However, research results also vary widely. A 2020 University of Maryland study suggests that phasing out coal plants has various costs: depending on climate and utilization scenarios, stranded assets could reach as high as US$127 billion (Cui *et al.*, 2020). Meanwhile, for lower-stranded assets valued between US$35 billion and US$9.3 billion, coal power plants will endure lost profits of US$51–64 billion from low loads (Cui *et al.*, 2020). Guangdong saw a significant loss in profit of over 300 billion RMB in an IPCC 1.5°C scenario. Another study states that China has US$158 billion at risk of becoming standard assets, especially with the current large amounts of coal capacity under construction and planning (Carbon Tracker, 2020).

For Guangdong, the costs of stranded assets depend on various scenarios and assumptions. If accounting for subcritical plants, stranded assets cost around US$16.4 billion. If using IPCC 1.5°C projections, 90% of China's coal plants will need to be retired by 2040, and stranded assets would be about US$44.01 billion (Sugden-Nalbandia, 2016). The overall costs of stranded assets decrease with higher discount rates and ages. The coal phase-out can be slowed down to reduce stranded assets via CO_2 capture and storage (CCS) (Cui *et al.*, 2020).

3. Oil Infrastructure

3.1. *Oil consumption*

Oil is the primary energy imported into Hong Kong, with HK$102 billion worth of oil products imported in 2019 (HKGOV, 2020). In terms of energy, this represents a 27% increase compared to 2014 (from 785 thousand terajoules to over a million), although the value of this import was 9% lower due to falling oil prices (HKGOV, 2020). A majority of oil products, such as fuel oil and aviation fuel, are sold to airplane and ship stores, respectively, with 99.6% of fuel oil (6.67 billion liters) going to ship stores and 99.9% of aviation fuel going to aircraft stores (HKGOV, 2020). Of the oil products imported into Hong Kong, only motor gasoline (94.7%, or 618.5 million liters) and kerosene (100%, or 3.168 million liters) are primarily sold locally. For gas oil, diesel oil, and naphtha, 43.9% (3.44 billion liters) are used in Hong Kong, and most of the rest, 50.4% (3.95 billion liters), go to ship stores (HKGOV, 2020). In total, the import of oil products was 24.135 billion liters of all types (including LPG) for 2019, of which 19.323 billion liters were reexported. Ship and aviation fuel comprised most of this exporting figure, with ship fuel (fuel oil, gas, oil, and diesel) at 10.65 billion liters going to ships passing through Hong Kong and a further 8.198 billion liters to aircraft. Compared to 2018, the consumption of aviation fuel, motor gasoline, LPG, and fuel oil all fell, while the usage of natural gas and gas oil rose (HKGOV, 2020). The trend is similar in value, although the value of motor gasoline increased while the amount imported decreased (HKGOV, 2019).

In Guangdong, the oil share of Scope 1 CO_2 emissions declined from 42% to 26% during 2000–2017 (Zhou *et al.*, 2018). Its share in the primary energy mix also fell from 35.0% to 25.9% over the same time frame (Guangdong Bureau of Statistics and National Bureau of Statistics, 2020). However, there was a slight rise in the composition of both oil and natural gas in the primary energy requirement energy mix during 2019–2020, from 25.9% to 27.2% and from 8.7% to 10.3%, respectively (Guangdong Bureau of Statistics and National Bureau of Statistics, 2022). However, Guangdong's energy consumption increased 3.66 times during that time, leading to a corresponding increase in crude oil consumption (226.6%) (Guangdong Bureau of Statistics and National Bureau of Statistics, 2020). The trend is much more apparent for household consumption, especially for vehicle fuels. In 2000, Guangdong households used 382 million liters of gasoline and 49 million liters of kerosene (Guangdong Bureau of Statistics and National Bureau of Statistics, 2020). By 2019, this had increased to 7.7 billion liters of gasoline and 210 million liters of kerosene, a 2034% and 428% increase, respectively (Guangdong Bureau of Statistics and National Bureau of Statistics, 2020).

3.2. *Oil transport infrastructures*

As all oil is imported in Hong Kong, most is received at one of seven oil terminals (MTMM (Hong Kong), 2020). These terminals generally accommodate ships of length 200 m or above and displacements of over 80,000 t (MTMM (Hong Kong), 2020). Further, the five terminals at Tsing Yi can accommodate 16 tankers simultaneously; some would be carrying vegetable oil. Ships over 195 m long must stay in the South Lamma Dangerous Goods Anchorage (MTMM (Hong Kong), 2020). The total docking (excluding restrictions) of all Hong Kong oil terminals was at a low estimate of 729.5 thousand tons (MTMM (Hong Kong), 2020). Tsing Yi accounts for 75.9% of the total oil docking facilities in Hong Kong.

Although China currently has 28 large oil pipelines operating and a further four under construction, only the 75 km 20.0 mta (million metric tons per annum) capacity Zhanjiang–Maoming oil pipeline and the 202 km 10 mta Zhanjiang–Beihai oil pipeline pass through Guangdong (Global Energy Monitor, 2020a). These two pipelines transport fuel from Zhanjiang port to the Maoming or Guangxi refinery (Global Energy Monitor, 2020a). Of the smaller pipelines, Guangdong has 43 crude oil lines totaling 729.7 km, transporting 51.05 million tons of oil, and 44 refined oil pipelines at 6,510.7 km in length, transporting 39.53 million tons in 2019 (Guangdong Bureau of Statistics and National Bureau of Statistics, 2020). Guangdong has a high one-time oil docking capacity of over 1.29 million tons between the Zhuhai, Guangzhou, Dongguan, Zhanjiang, and Shenzhen terminals, able to accommodate ships of up to 300,000 t (MTMM (Hong Kong), 2020; SINOPEC, 2020).

COVID-19 drastically increased the cost of charters within the oil market, especially during the drastic fall of oil prices in March–April 2020, with oil tankers being used as storage facilities (UNCTAD, 2020). The cost of renting a very large crude carrier (VLCC) between the Arabian Gulf and China increased the daily rental prices from US$18,300 to over US$128,000 by March and US$176,000 by April before falling to US$40,600 in June (UNCTAD, 2020). COVID-19 further highlighted the vulnerability of stranded assets, resulting in new oil tanker orders during 2020 falling to a 17-year low, with orders between January and July 2020 40% lower than in 2019 (Mohindru, 2020). Due to lower demand and the need for new emission controls, the value of all kinds of ships is falling, with high-priced tanker sales ending for now (Mohindru, 2020). The uncertainty around COVID-19 and oil's future in the energy mix by 2040 has significantly increased dealing in second-hand ships around the 15–20-year-old mark (Mohindru, 2020). This slowdown in new orders is interesting because the global growth of oil tanks by deadweight was already less than 1% between 2018 and 2019, and oil trade decreased by 1.1% between 2018 and 2019 (UNCTAD, 2019, 2020). An interesting side note is that even though orders are at a 17-year low, scrapping rates have also decreased drastically, with no VLCC being scrapped in the first seven months of 2020, in part due to the increased demand for oil storage as a result of low oil prices (Mohindru, 2020). These trends led to an increase of 5.8% in oil tanker deadweight tons between 2019 and 2020 (UNCTAD, 2020).

Overall, oil and gas carriers in 2020 saw a combined recycling of 2.3 million tons worth of shipping (2 million for oil and 280 thousand tons for gas) (UNCTAD, 2020). In terms of recycling ratio, this decreased by 71% for oil and 55% for gas (UNCTAD, 2020). The total asset value of oil tankers in China at the end of 2019 was US$13.28 billion, with a further US$7.24 billion owned through Hong Kong (UNCTAD, 2020). Hong Kong had the fifth largest merchant fleet, with 4.93% of the world's total, and mainland China held 11.15% of the world's total merchant fleet (UNCTAD, 2020). Accounting for all shipping assets (bulk carriers, passenger and container ships, chemicals, LNG, oil, etc.), Hong Kong held US$34.2 billion in ship assets and mainland China a further US$91.55 billion, with the value together being much higher than the total held by the first-placed Greece (US$96.8 billion) (UNCTAD, 2020).

3.3. *Oil storage infrastructures*

For Hong Kong, oil storage capacity estimates vary. The estimated storage capacity at Tsing Yi is 2.5 million barrels, or 369,000 m^3 (Reuters, 2010). However, other estimates suggest that ExxonMobil alone has 450,000 m^3 of storage, with Sinopec having a further 120,000 m^3 of storage (Tank Storage, 2009). This would be ignoring Shell and Chevron, who also have terminals in Tsing Yi. The Shell terminal was built in 1991 at a cost of HK$2.5 billion (Shell, 2019). Unfortunately, neither Shell nor Chevron provide data on the size of their storage systems at Tsing Yi. Estimates can be made using satellite images of the number and size of storage units at each site and known data on maximum ship size, which suggest that Shell sites have a capacity of around 75,000 m^3. In contrast, Chevron has larger terminals with much larger tanks, likely at a 150,000 m^3 capacity, while the PAFF terminal's capacity seems likely to be 80,000 m^3. Using these estimates, storage is expected to be around 875,000 m^3, or 5.93 million barrels. The total tank storage (of all locations in Hong Kong) is 936,436 m^3.

The Hong Kong government further subdivides the storage capacity (as of 2019) into the following sectors (HKGOV, 2020):

(1) *Aviation fuel*: fuel is stored at two locations; 12 storage tanks managed by the airport allow a storage of 220,00 m^3, while a further 264,000 m^3 (eight tanks) is located at Tuen Mun's PAFF terminal (Kong, 2021). The thoroughfare is six million tons a year, and fuel is transported from Tuen Mun to the airport via undersea pipelines (Towngas, 2021a).
(2) *Unleaded motor gasoline*: 101,282 m^3; 46,894 m^3 (46.3%) tank storage. Tank storage is enough for 22 days of use.
(3) *Kerosene*: 5,100 m^3 and tank storage is 2,014 m^3 (39.5%), enough for 260 days.

(4) *Gas oil, diesel oil, and naphtha*: 730,714 m^3, of which 380,271 m^3 (52%) is in tank storage, enough for 24 days of use.
(5) *Fuel oil*: 522,791 m^3, of which 223,659 m^3 is in tanks (42.8%), which is enough for 12 days.
(6) *LPG*: 17,516 t, with 6,833 t in tanks (39.3%), enough for seven days of use, and the remainder distributed throughout 67 LPG filling stations, although data on the size of each station are not available (HKGOV, 2018, 2020). LPG-only stations are run by ECO through Towngas, with five LPG filling stations supplying 30% (65,000 t) of Hong Kong's LPG vehicle market. Shell, Esso, PetroChina, and Towngas also provided LPG for vehicles and home use. PetroChina has five gas stations equipped with LPG, Esso has 12 stations with LPG (included in the totals above), and Shell sells LPG containers through its 39 storerooms.

Zhuhai in the Greater Bay Area is China's largest storage site for oil and gas, with 3 million m^3 of storage either in operation or under construction. Bonded storage is 1.17 million m^3, with Zhuhai Gaolan Port accounting for 25% of Guangdong's oil and 20% of its gas storage (Sofreight, 2020). The West Zhuhai storage terminal (22 tanks) has a capacity of 457,000 m^3 purely for oil products and LPG (Fortune Oil, 2020). Looking at satellite images and gathering information about the jetties' locations, the storage capacity of bonded storage sites at Zhuhai alone is most likely around 1 million m^3. If this 1 million m^3 figure is 20% of Guangdong's oil storage capacity, the total capacity would be 5 million m^3, or 4.4 million tons. The other large port facility is the Sinopec Zhanjiang port, with an oil storage capacity of 823,000 m^3 and a designed capacity of 45 million tons annually. Guangdong holds a strategic position within the global oil trade, with 30% (15 million bbl/d) passing through the South China Sea per day in 2016, of which 42% (6.3 million bbl/d) was bound for China (Dunn and Barden, 2018).

The lack of accurate information is expected as international experts give the current range of oil capacity in China's strategic petroleum reserve (12 locations) to be from 300 up to 600 million barrels at the end of 2019 (EIA, 2020). Estimates within China show similar numbers, suggesting that China's total oil storage capacity is 48.41 million tons, or 331 million barrels for total capacity (Zhao and Yutong, 2020). The article believes that the 70% (231.8 million barrels) threshold is the country's capacity limit, while as of July 2020, China had already reached 228.39 million barrels (Zhao and Yutong, 2020). However, including commercial stocks, China's crude inventories may have been as high as 831 million barrels in May 2020, with China adding 95.04 million barrels of storage worth in 2020 (Hellenic News, 2020). COVID-19 has also changed the storage situation, both for Guangdong and for China as a whole. After June 2020, China amassed 73 million barrels of oil in 59 different ships floating at sea, with oil brought in April and May building up offshore (Egan, 2020). This is seven times higher than the monthly average for the first quarter of 2020 (Egan, 2020). The most significant issue for Guangdong's oil infrastructure is bottlenecks in storage and transport, as shown by the number of large oil containers loitering off the Chinese coast in mid-2020 (Egan, 2020).

3.4. *Oil refineries*

Official statistics suggest that refinery processing capacity in Guangdong was 55.9 million tons in 2019, or 153.1 thousand tons per day (Guangdong Bureau of Statistics and National Bureau of Statistics, 2020). The list of oil refineries shows an average capacity of 1.124–1.224 million bbl/d (barrels a day), or around

150,000 t daily, including the Sinopec Maoming facility at 274,000 bbl/d, the Sinopec Guangzhou branch at 210,000 bbl/d, the Sinopec Dongxing plant at 100,000 bbl/d, CNOOC Huizhou at 240,000 bbl/d, and the Sinopec Guangdong Zhanjiang refinery project at 300,000–400,000 bbl/d. Refining capacity per year can be further subdivided into five sections: gasoline at 11.66 million tons, kerosene at 8.41 million tons, diesel oil at 14.7 million tons, fuel oil at 1.74 million tons, and LPG at 4.61 million tons (Guangdong Bureau of Statistics and National Bureau of Statistics, 2020).

Guangdong is expanding its capacity with two large oil refineries near completion, including the Zhongke refinery and petrochemical port, which began partial operation in July 2020. This Sinopec facility can process 200,000 bbl/d and will be integrated with the 100,000 bbl/d Dongxing plant for a total capacity of 300,000 bbl/d (*Business Times*, 2020), while the port can import 34 million tons per year, of which 5.61 million are refined oil, making it the largest such facility in China (NS Energy, 2020b). Moreover, the 400,000 bbl/d Venezuelan–China oil refinery in Jieyang in Guangdong was expected to begin operation in late 2021 (Reuters, 2018). These two refineries would increase Guangdong's refined oil processing capacity to 1.72 million bbl/d, or 251,630 t/day. Most of this capacity is from Sinopec, which runs about 1 million barrels daily in processing, and CNPC PetroChina, which is building the Jieyang refinery. The total refinery capacity in China was 860 million tons per year, meaning Guangdong controls 10.7% of the total refinery capacity in China (CNPC, 2020).

3.5. *Oil-fired power plants*

There is only one oil-fired power plant in the 278 MW heavy fuel oil plant in Fengsha, Guangdong (Global Energy Observator,

2011). Current information on whether the plant is still operational is not available. Simultaneously, the United States's Energy Information Administration (EIA) reports that in 2018, less than 1% of China's installed capacity was from oil (EIA, 2020). Hong Kong does have some oil-fired units at Lamma Power Station, with four 125 MW units and one 55 MW unit (555 MW) (HK Electric, 2014). Macau's Coloane Power Station A also uses oil products, primarily diesel and fuel oil, in its eight-unit generating capacity of 271.4 MW (CEM, 2020b).

4. Natural Gas Infrastructure

4.1. *Natural gas consumption and supply*

Hong Kong has planned to increase natural gas's share in its energy mix from 27% in 2015 to 50% by 2020 to allow the phasing out of coal in electricity generation (HKGOV LEGCO, 2020) (HKGOV, 2017b). Doing so helped it achieve a 50–60% reduction in carbon intensity and a 20% absolute drop in carbon emissions compared to 2005 levels (HKGOV LEGCO, 2020; HKGOV, 2017b). A 7% increase in natural gas imports occurred between 2009 and 2019, from 2.27 million tons to 2.44 million tons, with all imports from mainland China (HKGOV, 2020). During the same time, the value of gas increased from HK$4 billion to HK$10.47 billion, denoting a change in unit price of gas from HK$1.77/kg in 2009 to HK$4.29/kg in 2019 (HKGOV, 2020).

In Guangdong, natural gas accounts for 8.3% of energy consumption as of 2018, increasing from 0.2% in 2004 (Guangdong Bureau of Statistics and National Bureau of Statistics, 2020). LNG usage in China and globally is expected to increase, with global demand expected to reach 400 million tons by year-end 2022, up 6.6% from 2021, and China and other emerging markets accounting for most of the expanding demand (Bloomberg, 2022). China is significantly increasing the number of ships using LNG as a fuel source due to its much lower NO_x and SO_x emissions than heavy fuel oil (SNAM, 2020). According to EIA, China increased its production of gas by 8% between 2018 and 2019, while offshore (South China Sea) natural gas production rose by 10.4% (EIA, 2020). This occurred as the consumption of natural gas between 2009 and 2019, on average, increased by about 13% per year (EIA, 2020). Data from CNOOC suggest that in 2019, its production of natural gas in China was 987.9 million cubic feet per day, of which 87.8% was from the South China Sea (CNOOC, 2019). Total reserves were estimated at 6.36 trillion cubic feet in 2019, of which 17.4% were located in Bohai and 81.4% were from the South China Sea (CNOOC, 2019). CCS in the form of EOR within southern China could also generate revenue of US$2 billion using oil prices of US$50 per barrel (Wang *et al.*, 2020).

China is the largest natural gas importer, importing 130.3 billion m^3 of gas (consumption was 305.85 billion m^3) in 2019. CNPC sold over half that amount, with its Chinese natural gas sales at 181.3 billion m^3 in 2019 (CNPC, 2020; EIA, 2020). China's total gas usage is also expected to increase to 360 billion m^3 by 2025 (Liang and Ang, 2021). Longer projections suggest a doubling in demand by 2040, or a 3.9% annual increase, with China making up 12% of the global market by 2040 (SNAM, 2020). Imports also increased during 2020 and are expected to increase further (Liang and Ang, 2021). Finally, CNPC expects pipeline gas imports to reach 100 billion m^3 by 2025 and LNG capacity to be 48% higher than current levels at 120 mta (Liang and Ang, 2021). Guangdong's yearly import capacity via pipelines alone was 117.7 billion m^3 of gas, including the 90 billion m^3 capacity of the West–East Gas Line 2-3 (EIA, 2020; Global Energy Monitor, 2020a). For larger pipelines solely in southern China, including the 10 billion m^3 Guangzhou–Nanning branch line, Guangdong has 27.7 billion m^3, or 20.36 million tons a year, in pipeline capacity, with 14.3 billion m^3 (10.51 million tons) dedicated solely to LNG transport (Table 1). This does not include the recently built Shenzhen LNG line linking the Shenzhen and Daifu LNG terminals (Global Energy Monitor, 2020a).

International LNG supply for China primarily comes from Australia (46% of LNG) and Qatar (14.5%), with LNG making up 62% of China's natural gas imports in 2019 (EIA, 2020). Demand remained high even with COVID-19 compared to that in other regional importers, with Chinese gas imports up 4.8% in Q2 of 2020 compared to 8.3% and 8.9% drops in Japan and Korea, respectively (IGU, 2020). As such, if production can be increased, no significant additional investment in infrastructure would be necessary. CNOOC, China's primary offshore oil and gas company, agreed that in 2019, the natural gas supply was relatively abundant (CNOOC, 2020). However, countrywide production was only about half of the consumption (EIA, 2020). This shift to gas has been seen as one way to decrease China's CO_2 emissions, but data on natural gas' CO_2 emissions often ignore or downplay the severity of methane leakages during transport and production (Lydia *et al.*, 2020). Some studies have suggested that its overall greenhouse gas emissions would be the same, if not slightly higher, after accounting for methane leakages compared to coal (Lydia *et al.*, 2020).

In 2020, there was a slight opening in the Chinese gas and oil market, with restrictions on foreign investments being removed (Wang *et al.*, 2020). The new resource tax law in September 2020 eliminates the tax associated with oil and gas used during their heating and extraction, and it offers 20–30% tax reductions for other oil and gas extraction processes to encourage enterprises' development in the oil and gas sector (Wang *et al.*, 2020). Pipe China was established in December 2019 and will slowly take over transport and distribution from the current state-owned

Table 1. China natural gas import routes.

Transport Mode	Routes	Length (km)	Capacity (bcm/year)	Completion Year
Inland transport routes	Central Asia – China Gas Pipeline A	1,833	15	2009
	Pipeline B	1,833	15	2009
	Pipeline C	1,830	25	2012
	Pipeline C	1,000	30	2014
	Myanmar–China gas pipeline	2,520	12	2012
	Russia–China gas pipeline	3,968	38	2019
Seaborne shipping routes	The Middle East/North Africa – Indian Ocean – Malacca Strait – South China Sea – China	9,405 (Doha to Shanghai	N/A	N/A
	Australia/Southeast Asia– South China Sea – China	3,900 (Darwin to Shenzhen	N/A	N/A

oil and gas giants. Its primary goal is to improve resource allocation and safeguard energy security (Wang *et al.*, 2020).

4.2. *Natural gas pipelines*

Gas is imported directly via submarine pipelines from Mainland China to Hong Kong's Black Point, Castle Peak, and Lamma power stations for electricity generation. Simultaneously, some gas is sent to the Tai Po Plant by the company Towngas to produce town gas (HKGOV, 2017a). Black Point receives its gas directly from CNOOC's offshore gas fields and, in extreme circumstances, from the Gaolan terminal (CNOOC ESG, 2020). Natural gas for town gas production (with feedstocks being 38% naphtha and 61% natural gas) is transported to Hong Kong via a 34 km twin submarine pipeline from the Shenzhen Dapeng Bay LNG terminal (EMSD, 2015). Town gas is produced at two locations within Hong Kong, with the Tai Po Plant producing the majority and the Ma Tau Kok Plant producing the rest. The maximum capacity of these two plants is 12.6 million m^3 a day (Towngas, 2019). Tai Po has a capacity of 9.7 million m^3, and Ma Tau has a capacity of 2.6 million m^3. Town gas is transported via a 3,500 km pipeline network (HKGOV, 2017a). Usage of town gas stayed relatively steady between 2018 and 2019, with a drop of 2.8%, although the number of customers increased by 1.3% to 1.93 million (Towngas, 2020b). Natural gas production in 2020 released 345,000 t of CO_2, equivalent to a carbon intensity of 0.592 kg of CO_2 per unit of town gas, or a 23% reduction compared to 2005 (Towngas, 2021b). Towngas' assets in property, plants, and equipment were HK$61.08 billion at the end of 2019 (Towngas, 2020b). Total natural gas consumption in 2019 for Hong Kong was 2.44 million tons (3.3 billion m^3) (HKGOV, 2020), of which CNPC provided slightly under half (1.5 billion m^3) and CNOOC supplied CLP's Black Point natural gas power station (3.17 GW) (CNOOC, 2019; CNPC, 2020).

Guangdong holds a significant share of the current importing and planned capacity for the next few years. However, 2019 import data show that the current import rate is well below capacity, with Guangdong only importing 4.83 million tons of foreign natural gas, all of which was LNG in 2019. After entering China, all piped natural gas would be considered domestic since it had already passed the borders (Guangdong Bureau of Statistics and National Bureau of Statistics, 2020). Total natural gas production in 2019 was 11.2 billion m^3, while consumption was 20.1 billion m^3 (Guangdong Bureau of Statistics and National Bureau of Statistics, 2020). Most of the rest were supplied by CNOOC and CNPC (CNOOC, 2020). One major supplier, CNPC, controls 75.1% of China's total natural gas pipelines and supplied 4.6 billion m^3 of natural gas in 2019 to Guangdong (CNPC, 2020).

International gas imports from Turkmenistan provide 65.8% of China's international piped gas imports (31.7 billion m^3) (EIA, 2020), while LNG is by far the most significant vector for imports, accounting for 62% (78.5 billion m^3) of China's natural gas imports (EIA, 2020). For imports, in 2019, Australia was the largest country of origin at 28.2 million tons, followed by Qatar (7.5 million tons) and Malaysia (7.5 million tons) (IGU, 2020). Newly built pipelines are expected to increase these import figures, with Line D of the Central Asia–China Pipeline system increasing capacity from Turkmenistan to 65.1 billion m^3 (Table 1) (EIA, 2020). Meanwhile, the China–Russia Pipeline will provide up to 38 billion m^3 of gas by 2023 (Towngas, 2020a). Russia will likely continue to be a reliable supplier for China in the gas sector, with CNPC signing a 30-year US$400 billion deal for the China–Russia Pipeline in 2014 (Zhang and Bai, 2020). Volume amounts increased to 10 bcm in 2021, 15 bcm in 2022 and expectedly 22 bcm for 2023 (Enerdata, 2023). Additional planned construction of the pipeline is expected to increase total capacity to 38 bcm/year by 2024 (Enerdata, 2023). Although this has diversified China's gas supply, most of the effect will be limited to northern China, which is unlikely to affect Guangdong significantly (Towngas, 2019; EIA, 2020). Overall, China is expanding its gas network from an estimated 104,000 km in 2020 to 163,000 km in 2025, with 22 of 31 provinces committed to significantly expanding its network (SNAM, 2020). For pipelines, underground storage tends to be preferred due to its larger capacity, with the Sinopec Wen 23 storage facility in Henan alone having over 238 million m^3 of storage space (SINOPEC, 2020).

In 2018, China became the world's largest natural gas importer and second-largest LNG importer, and it grew be the largest LNG importer in 2021 (George *et al.*, 2020; Energy Institute, 2023). Imports account for 45% of China's gas consumption (George *et al.*, 2020). However, China has significant issues with non-LNG storage capacity, with total storage at 11.8 billion m^3 in 2018, less than 5% of total gas consumed, while the United States and Russia can store up to 20% of annual gas consumption (Zhao and Lawson, 2020). China aims to increase this to 10% of demand (SNAM, 2020). Some companies, such as Towngas, have recently opened extensive storage sites, including the construction of the Jintan (Jiangsu) storage sites that operate 25 wells with 1.1 billion m^3 of storage capacity (Towngas, 2020b). Regarding offshore extraction and pricing, CNOOC (China's largest offshore oil and gas producer) in 2019 found nine new fields in the South China Sea and appraised an additional 22 fields within the area (CNOOC, 2019). The new (December 2019) Lingshui 17-2 field alone adds a maximum capacity of 3 billion m^3 annually to Guangdong, Hong Kong, and Hainan. The price in 2019 was US$6.27 per million cubic feet, down by 2.2% from 2018 (CNOOC, 2019). Total production in the South China Sea was at 866.7 million cubic feet per day, or around 8.95 billion m^3 per year.

4.3. *LNG infrastructures*

Tracking gas infrastructure, Global Energy Monitor found 55 LNG terminals and 66 gas pipelines in operation or being constructed within China as of July 2020 (Global Energy Monitor, 2020a). In April 2019, China had 21 LNG terminals under construction with an annual capacity of over 80 million tons (George *et al.*, 2020). Nine terminals are coming online in 2022, with four existing terminals expanding for 38.9 million extra tons of capacity (Bloomberg, 2022). Total import capacity is expected to increase to 93.5 million tons by 2027, with utilization rates dropping to around 60% in 2022 (Bloomberg, 2022). While natural gas is mainly used in electricity generation and heating, it can also be used in transport. Studies done in Zhejiang Province show a lower cost per 100 km with LNG than with diesel in almost every month from September 2013 to September 2018. Six pipelines begin, end, or cross Guangdong, and eight LNG terminals are located in Guangdong, while a further eight LNG terminals and two gas pipelines are being constructed within the province (Global Energy Monitor, 2020a). If including smaller pipelines, Guangdong has 52 gas pipelines with a length of 1947.1 km and a total pipeline traffic of 8.86 million tons, up from 3.77 million tons in 2000 (Guangdong Bureau of Statistics and National Bureau of Statistics, 2020). Data from household consumption also show that LNG use increased by 80% between 2000 and 2020 (Global Energy Monitor, 2020a). China has been expanding rapidly in the LNG market since 2020, having only 9% of the global import potential in 2020 (Plante *et al.,* 2020). However, in the same year, China had 37.7% of global import LNG terminals in construction and 40.5% of proposed LNG import terminals (Lydia *et al.*, 2020).

Furthermore, five out of the top six global LNG import capacity owners are Chinese companies (Lydia *et al.*, 2020). As of May

2020, China's LNG import capacity was 76 million tons (103.3 billion m^3), with a further 54.2 million tons under construction (73.7 billion m^3) (Lydia *et al.*, 2020). China's actual imports in 2019 were 61.7 million tons (81.8% of capacity), up by 7.7 million tons from 2018 (IGU, 2020). This increase in LNG traffic was not limited to China, with LNG trade increasing by 8.9% during 2017–2018, from 292 million tons in 2017 to 318 million tons in 2018, compared to a 0.6% increase in crude oil trade (UNCTAD, 2019). In Guangdong, the import capacity of current LNG terminals is 20.7 mta (28.14 billion m^3), and it would reach 36.7 mta (49.9 billion m^3) with planned terminals (Global Energy Monitor, 2020a). The new 1.1 billion m^3 Shenzhen gas terminal opened in 2019 (SNAM, 2020).

LNG vessels: As China continues to increase its LNG capacity, the cost of new LNG carriers must also be considered. LNG transportation costs 25–30% of LNG's total value and 10–30% of an LNG project's price (Kamalinejad *et al.*, 2016). Demand is high, with 281 LNG ships delivered and a further 95 on order in 2009, while capacity wise the global fleet capacity was expected to rise from 60 million m^3 in 2015 to 83.4 million m^3 by 2019 through the expected operation of 410 vessels (Kamalinejad *et al.*, 2016; Raju *et al.*, 2016). Actual growth was faster than predictions, with 541 vessels at year-end 2019, including 34 storage regasification units and 126 ships on order (IGU, 2020). However, the cost of LNG carriers is high, with large 215,000 m^3 LNG carriers costing up to US$250 million in 2014 and up to US$290 million for 270,000 m^3 ships in 2004–2005 prices (Kamalinejad *et al.*, 2016; Fikri *et al.*, 2018). Even smaller ships with a capacity of around 135,000 m^3 can have an initial capital cost of US$170 million (Kamalinejad *et al.*, 2016; Fikri *et al.*, 2018).

LNG carriers' average size increased, with 31% of orders in 2019 being for the size class of 160,000–173,000 m^3, while only

15% of active LNG ships in 2015 were of that size class (Raju *et al.*, 2016). Although the cost per m^3 capacity does decrease as ship size increases (for example, 6000 m^3 = US$45–55 million, 12,000 m^3 = US$50 million, and 30,000 m^3 = US$105 million) (Fikri *et al.*, 2018), this will still require substantial investment and lead to the possibility of significant stranded assets if natural gas were to be phased down for carbon neutrality. At the end of 2019, China's LNG fleet had an estimated value of US$4.27 billion. Hong Kong had an additional US$1.18 billion in LNG carriers; this number has likely expanded due to the LNG market growth over 2020 (UNCTAD, 2020). Further, for investors hoping to avoid the high capital cost, the demand for LNG in Southeast Asia has drastically increased rental prices, with prices up to US$350,000 per day (Miller, 2021). Average rental rates in early January 2021 were over double (US$215,000) those in January 2020 (US$103,000) (Miller, 2021). LNG vessel owners must be cautious due to high market volatility, especially shipbuilding costs. Further restrictions may be placed on LNG carriers as annual CO_2 emissions per ship are incredibly high at over 70,000 t, 0.7% lower than cruise ships and much higher than tankers (less than 20,000 t annually) (UNCTAD, 2020).

4.4. *Natural gas storage*

Of the five current and three under-construction LNG terminals in Guangdong, five have data on storage, with a total capacity of 3.04 million m^3 (1.307 million tons) of LNG, including two 480,000 m^3, two 640,000 m^3, and one 800,000 m^3 facilities (Global Energy Monitor, 2020a; Industry About, 2020).[1] Specifically, larger LNG terminals, such as the Dapeng LNG

[1] The concentration of natural gas in LNG is much higher than in the gas form, with 1 billion m^3 of gas equaling 860,000 thousand tons. As such, LNG has a density of 430 kg/m^3, much higher than pure natural gas (680 g/m^3), which has a density of about half that of air (1.293 kg/m^3).

terminal (7.7 mta), serve dual purposes, with LNG importing capacity from ship sources via a pipeline through the Trunkline pipeline project (Global Energy Monitor, 2020a). Further, both are currently operating and under construction, and plants are seeing expansions through significant investment. For example, the Huaying Chaozhou Guangdong plant (operational 2023) is seeing 8 billion RMB investment in its first phase (storage will be 600,000 m^3), an additional 4 billion RMB for ancillary facilities in the terminal, and a further 2.3 billion for an LNG carrier (Reuter, 2020). The second phase saw an additional 800,000 m^3 in storage (Reuter, 2020). While large state-owned corporations have traditionally run LNG plants, especially CNOOC, there has been an expansion in private players in both the pipeline and the LNG market (Zhao and Lawson, 2020).

4.5. *Natural gas-fired power plants*

Under the CLP's Five-Year Plan (2018–2023), the company plans to spend HK$52.9 billion on new capital investment, including the construction of a new gas-fired generation unit at Black Point and a floating LNG terminal near Lantau, which would also provide natural gas to Lamma Power Station (CLP, 2020b). During 2014–2019, CLP's total carbon intensity per kWh decreased from 0.64 kg CO_2 to 0.50 kg CO_2, with a goal of 0.4 kg CO_2 for 2020 with the new gas-fired units (CLP, 2020b). HK Electric at Lamma Power Station had a natural gas capacity of 680 MW, and as mentioned earlier, it is slowly increasing (HK Electric, 2014). Three new gas-fired units were commissioned in February 2020 (HK Electric, 2020b). With the L10 generator, Lamma changed its fuel mix to 50% coal and 50% gas from 70% coal and 30% gas in 2019, while with the L11 generator, it should be 55% gas in 2022 and with the L12 generator, 70% gas by 2023 (Figure 1) (HK Electric, 2020a).

The total asset value of all land, fixed assets, and rights of assets at the end of 2019 for CLP was HK$150 billion (CLP, 2020a).

CLP's annual report data suggest that the value of transmission line assets was likely just under HK$80 billion in 2019 (CLP, 2020a). The construction cost for Black Point alone was estimated at around HK$24 billion (Power Technology, 2020). However, as all Hong Kong-based CLP plants (Castle Peak, Black Point, and Penny's Bay) are run by Castle Peak Power Company (CAPCO) via a 70/30 equity split between CLP and China Southern Power Grid (CSG), respectively, it is possible to make some deductions (CLP, 2020a). CLP seemed to have a non-current asset value of HK$34.2 billion in CAPCO in 2019; hence, the total CAPCO asset value would likely be around HK$48.9 billion (CLP, 2020a). Revenue from CAPCO was HK$16.8 billion in 2019 for a profit of HK$2.7 billion (CLP, 2020a). For Hong Kong electric plants, machinery had a value of HK$54.43 billion, with a further HK$9.8 billion in construction assets (HK Electric, 2020a). Total property, plant, and equipment were valued at HK$66.6 billion in 2019, including the joint-venture floating LNG terminal with CLP (HK Electric, 2020a). According to the 2020 Hong Kong Electric annual report, total assets accumulated were HK$111.5 billion at year-end 2020, up from HK$109.7 billion in 2019 (HKEI, 2022).

Macau has two main local power stations. The smaller one has three generators (Coloane Power Station B), with a total capacity of 136.4 MW. Gas is imported directly from Mainland China via an underground pipeline (CEM, 2020b). Gas usage fluctuates significantly yearly, with consumption peaking at 178 million m^3 in 2017 and 136 million m^3 in 2019 (Government, 2020).

Natural gas-fired power generation capacity in Guangdong at the end of 2020 was just over 20 GW, higher than any other province (Figure 1 in Chapter 3) (The Oxford Institute for Energy Studies, 2020). Under Guangdong's 2021–2025 plans, it hopes to increase its capacity to 42 GW to decrease coal usage (Figure 2 in Chapter 3) (The Oxford Institute for Energy Studies, 2020). However, the high cost of turbines, especially with the new

H-class, can cause problems, as turbine costs are 30% higher than coal plant generators of the same class (The Oxford Institute for Energy Studies, 2020). The increase in gas is projected in the 13th Five-Year Plan for Natural Gas Development, which called for a rise in China's use from 5.9% of primary energy consumption in 2015 to 10% by 2020 (Zhao and Lawson, 2020).

5. Assets of Fossil-Fuel Infrastructures

According to the current energy infrastructure development plans with project-level information, the Greater Bay Area (GBA), including Hong Kong and GBA Guangdong, is transitioning from coal to natural gas for local electricity generation (Figures 1 and 2). Hong Kong is replacing its retiring coal-fired power plants with natural gas-fired ones, and the transition is expected to be completed within one decade (Figure 1). For GBA Guangdong, coal-fired power capacities have remained stable for 15 years until around 2025, while natural gas-fired power capacities are increasing rapidly from 13,970 MW in 2015 to 25,047 MW in 2020 and 47,707 MW in 2025, which shifts the fuel mix of electricity generation from coal to the much less carbon-intensive natural gas (Figure 2). The shares of coal- and natural gas-fired power plants among total local generation capacities in GBA Guangdong were 49.2% and 27.1% in 2015, 37.3% and 36.0% in 2020, and 26.3% and 47.2% in 2025 (Figure 2). The energy transition is expected to continue in the coming decades, with more planned natural gas-fired power capacities coming online.

In stark contrast, the much poorer non-GBA Guangdong has far fewer natural gas-fired power plants at present and in plans, being 75 MW in 2020 and 3,420 MW in 2025, while coal-fired power capacities were 39,778 MW in 2020 and 44,978 MW in 2025. Many non-fossil electricity generation capacities — including

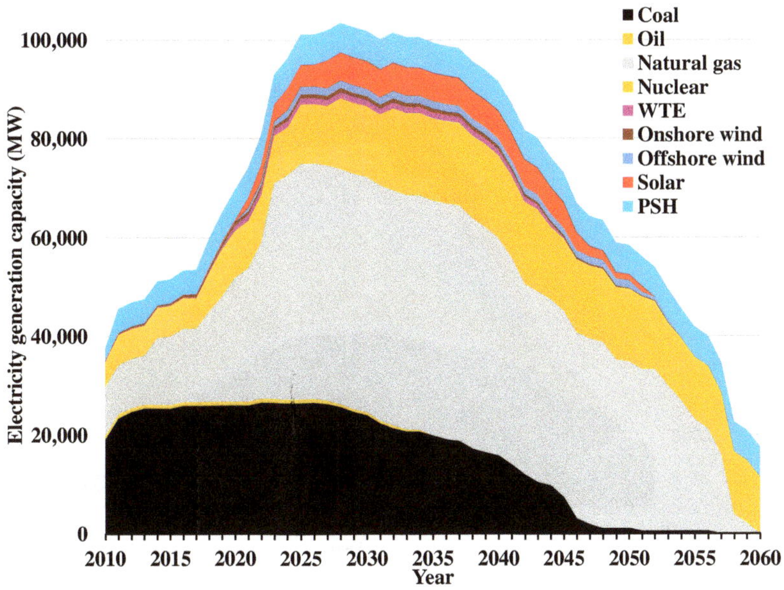

Figure 2. *Electricity generation capacity in GBA Guangdong* (numbers after 2022 include those planned and retiring power plants).

nuclear and, most importantly, offshore wind and solar — have been planned to witness a very different energy transition (Figure 3), especially with the expected retirement of current coal-fired power plants. Fossil-fuel-fired electricity generation capacities are primarily in GBA rather than the less developed non-GBA Guangdong. In 2020, the shares of GBA Guangdong, Hong Kong, Macau, and non-GBA Guangdong were 51.1%, 9.5%, 0.1%, and 39.3%, respectively (Figure 4). According to the current plan, GBA will still host most fossil-fuel-fired power plants.

Energy transition is not identical to asset transition. The latter could come faster than the former because the asset value of existing and planned fossil-fuel infrastructures will depreciate over time. We consider the average lifetime of various energy

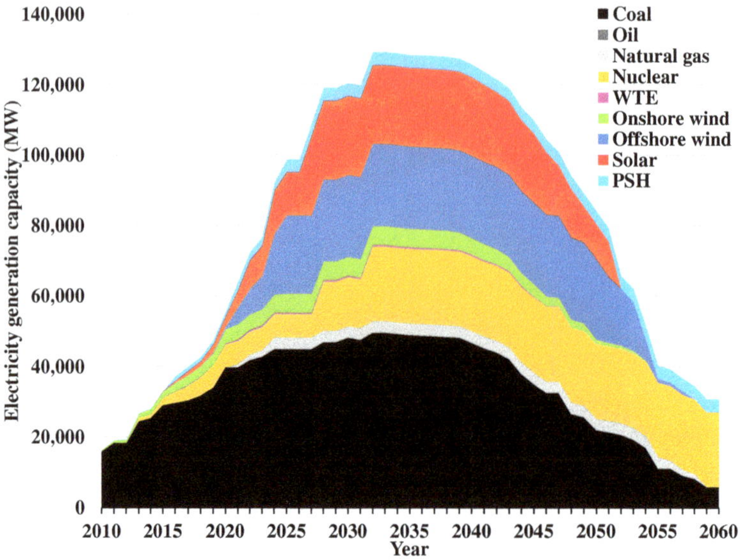

Figure 3. Electricity generation capacity in non-GBA Guangdong (numbers after 2022 include those planned and retiring power plants).

infrastructures (Table 2), and the depreciation rate is estimated to be linear. A caveat is that not all power plants or energy infrastructures are closed exactly when their designated lifetime is reached. Some may retire prematurely, and some others may have their services extended. Nevertheless, the analysis provides helpful insights into the evolution of energy transition and asset values before 2050 and 2060, the target years of Hong Kong and entire China for reaching carbon neutrality, respectively.

With the continuous depreciation and retirement of coal-fired power plants in Hong Kong, natural gas-fired power plants have become dominant. The newly completed floating LNG terminal also adds valuable assets. Our estimate shows that existing and planned fossil-fuel infrastructures will see their asset values depreciating by about 80% in 2050 and to a very low level in 2060. GBA Guangdong also witnessed a steady decline in the

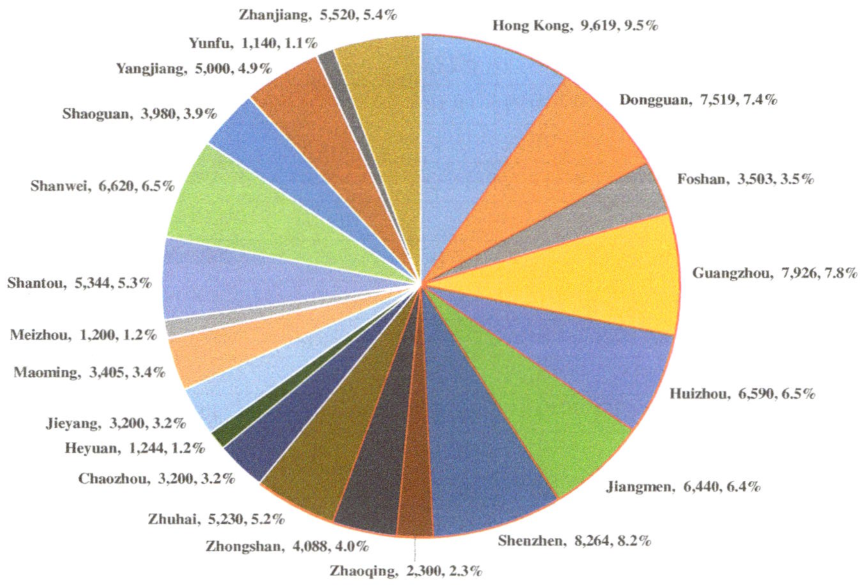

Figure 4. *Fossil-fuel-fired electricity generation capacity (MW) in 2020 for Guangdong by municipalities* (red boundaries indicate those in the GBA).

Table 2. Critical parameters in calculating the asset values of energy infrastructures.

Assets and Fuel/Type		Capital Costs		Lifetime (Years)	Capacity Factor
Electricity generation	Coal	0.66	Million USD/MW	35	4448
	With CCS	104% increase			
	Natural gas	0.56	Million USD/MW	35	2614
	With CCS	139% increase			
	Nuclear	2.50	Million USD/MW	50	7802
	Offshore wind	1.10	Million USD/MW	30	3066
	Onshore wind	0.76	Million USD/MW	30	2232
	Solar	0.44	Million USD/MW	25	1281
	Waste to Energy (WTE)	3.06	Million USD/MW	25	5563
Energy storage	Pumped Storage Hydro (PSH)	0.66	Million USD/MW	80	
	Battery storage	0.30	Million USD/MWh	15	
Electricity transmission	Transmission lines			80	
Natural gas	LNG terminal	352	Million USD/million ton	50	
	Gas pipeline	1.70	Million USD/km	30	
Oil	Oil terminal	0.81	Million USD/1000 berth tons	30	
	Refinery	22.4	Million USD/Thousand barrels per day	50	

Note: Unit capital costs are taken from the International Renewable Energy Agency (IRENA), Global Energy Monitor, and China's domestic sources. The values are optimistic estimates because our calculations are stretched from 2010 to 2060 without considering variations and learning effects. Capacity factors are taken as China's national average values in 2021 (China Electricity Council, 2022). The cost increase ratios of CCS for coal- and natural gas-fired power plants are calculated (Zero Emissions Platform, 2011).

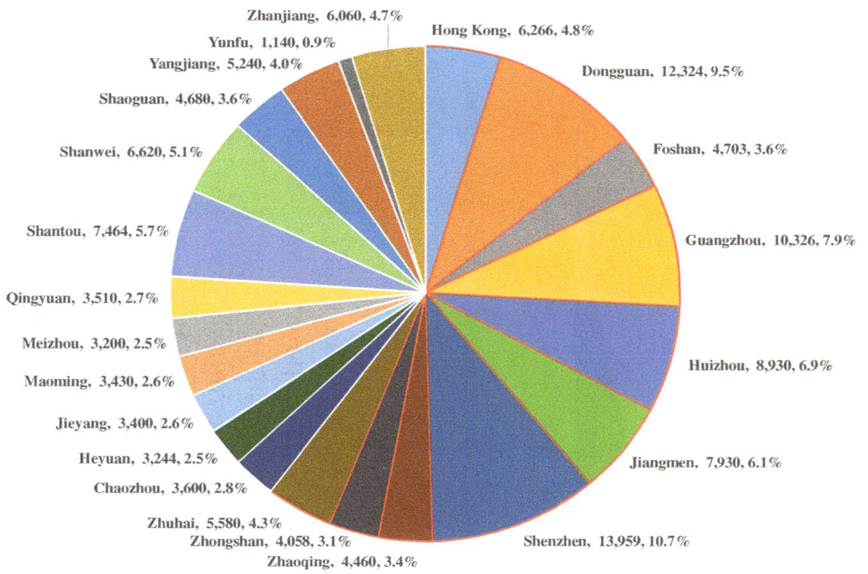

Figure 5. *Fossil-fuel-fired electricity generation capacity (MW) in 2030 by municipalities* (including those planned and retiring power plants; red boundaries indicate those in the GBA).

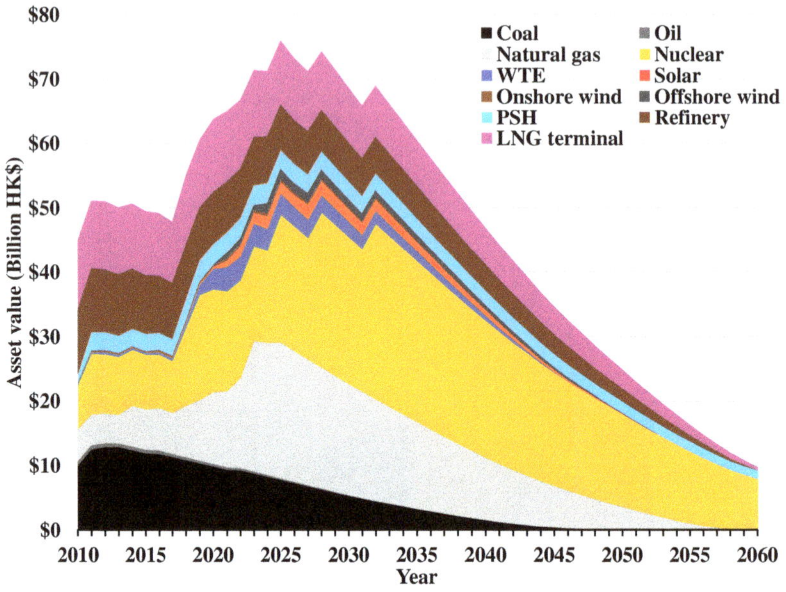

Figure 6. *Energy infrastructure assets in GBA Guangdong* (numbers after 2022 include those planned and retiring).

asset values of coal-fired power plants. In contrast, new investments in natural gas-fired power plants keep increasing their asset values despite depreciation (Figures 4 and 5). Together with oil refineries and LNG terminals, fossil-fuel infrastructures are crucial components of total assets in the 2020s, peaking at US$47.2 billion in 2023 and then declining to US$8.6 billion in 2050 and US$0.5 in 2060 (Figure 6).

Non-GBA Guangdong also has heavy assets of fossil-fuel infrastructures, but the structure is highly different from that of GBA Guangdong. With the depreciation of coal-fired power plants and few planned fossil-fuel-fired electricity generation capacities, fossil-fuel assets in the power sector are expected to diminish from US$21.1 billion in 2023 to US$3.7 billion in 2050 and US$0.6 billion in 2060 (Figure 7). LNG terminals and oil

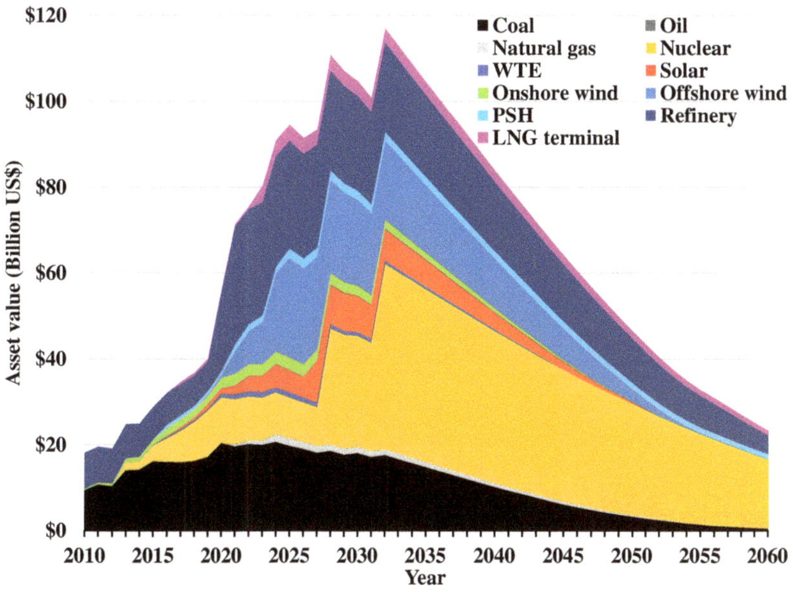

Figure 7. *Energy infrastructure assets in non-GBA Guangdong* (numbers after 2022 include those planned and retiring).

refineries will become more critical, going from US$30.1 billion in 2023 to US$12.1 billion in 2050 and US$5.4 billion in 2060 (Figure 7). Oil refineries will occupy a great majority of fossil-fuel assets in 2060. As a result, they could be the most critical stranded assets if carbon neutrality in Guangdong mandates their premature retirement or phasing down.

Overall, the potential concerns about stranded fossil-fuel assets under carbon neutrality in 2060 or even 2050 will not mainly be about existing projects in the pipeline but mostly about future choices of energy infrastructures.

Chapter 3

Non-Fossil Fuel Infrastructures

1. Global Trends

We have witnessed a significant rise in the use of renewables worldwide, with an increase of over 230 GW of renewable capacity in the power sector during 2020–2021, of which 117 GW was in China (REN 21, 2021). Solar and wind led the way during COVID-19, with additions of 164.3 GW and 93 GW of capacity globally in 2020 alone (REN 21, 2021). China was also the most prominent overseas renewable financier in nine of the ten years between 2009 and 2019 (Charlie, 2019). Chinese funding prior to COVID-19 was mainly on the rise, aiming to help facilitate goals set out in the Belt & Road Initiative (BRI), which gave below-market rate loans for renewable energy projects (Charlie, 2019). The contribution of renewables is expected to increase significantly, from 28% of the current electricity supply to 43% by 2030 under the currently stated policies. This increase is driven by solar energy, growing from 4% to 12%, and wind energy, growing from 7% to 13% (IEA, 2022).

Increased demand for renewables has led to a premium on dispatchable energy systems, such as hydropower, which can be ramped up quickly (IEA, 2022). Hydropower is seen as an essential aspect to make up for supply-side gaps in a primarily renewable electric system (IEA, 2022). As such, hydropower has seen significant global growth in utilization under all stated, announced, and net-zero goals for 2030 and 2050 (IEA, 2022). However, hydropower is sensitive to changing weather patterns, especially those associated with reducing water availability, as seen in Brazil , Southern Europe, North Africa, and the Middle East over 2022 (IEA, 2022). For other power sources such as Geothermal usage increased by about 40% between 2010 and 2021. However even under the stated policy scenarios, with an impressive annual growth rate of 7.2% from 2021 to 2030, geothermal will only amount to 1% of the global electricity supply.

Globally, in 2022, 437 nuclear reactors were operational (393.6 GW), with an additional 60 under construction (61.9 GW), providing about 10% of global electricity (World Nuclear Association, 2022). Nuclear electricity production has remained relatively stagnant since the early 2000s, with minor changes between years (World Nuclear Association, 2022). In terms of the proportion of nuclear in the total production, the United States ranks first, with 19.6% of the country's electricity generated from nuclear energy, or 771.6 TWh, followed by China's 383.2 TWh and France's 363.4 TWh (World Nuclear Association, 2022). With regard to construction and grid connections, 2022 saw six reactors come online (in China, Finland, South Korea, the UAE, and Pakistan), with a further eight beginning construction (in China, Egypt, and Turkey), while four reactors were shut down (three in the UK and one in the US) (World Nuclear Association, 2022). Globally, in comprehensive terms, construction tends to dominate within Asia, especially China, India, and Russia, while expected nuclear phaseouts are mainly in Western Europe, especially in Germany (World Nuclear Association, 2022).

2. Non-Fossil-Fuel-Fired Power Plants

2.1. *Nuclear*

Since 1994, Hong Kong has been receiving electricity from the nearly 2,000 MW Daya Bay Nuclear Power Station. The two-reactor Daya Bay Nuclear Plant (built in 1993), located about 50 km from Hong Kong (in Shenzhen), currently sends 80% of its electricity supply to Hong Kong, with plans to continue to do so until at least 2034 (CLP, 2016). About 25% of Hong Kong's electricity is produced from this plant (HKGOV, 2017). One possible plan is to increase Hong Kong's nuclear share to 50% of total power (World Nuclear Association, 2020). China General Nuclear Power Corporation (CGN) expects to contribute significantly to this plan as the majority shareholder in Daya Bay (CGN, 2020a).

Data from Macau are inconclusive, as less than 14% of the energy used by the city is produced in Macau, with the other 86.33% imported from China Southern Power Grid (CSG) (CEM, 2020a). CSG shows the amount of nuclear energy in their mix at 4.15%, which indicates that Macau receives 3.6% of its electricity from nuclear energy (CSG, 2019). Ten reactors are located in Guangdong Province, providing 18% of electricity and 10% of the installed capacity (Zhou *et al.*, 2018c). At the end of 2016, Guangdong installed 10.47 GW of nuclear (Liu *et al.*, 2020). This figure has likely increased, as China had over 30 GW of nuclear generation capacity under construction in 2016 (CCP, 2016). A further 41 reactors (46 GW) are planned, while an additional 78 reactors have been proposed (106.5 GW) (World Nuclear Association, 2020). Government information shows 65.93 GW of nuclear in operation or under construction, the second-highest globally and the highest amount under construction (China Daily Global, 2020). Studies conducted by the government on how China can limit climate change to 1.5°C also state that China must increase its nuclear-generating capacity to 554 GW by 2050, a 28% share of the expected energy mix requiring at least 8.7 trillion RMB in investment (World Nuclear Association, 2020).

According to CGN, the nuclear power operator in Guangdong is building 16.1 GW of nuclear power (CGN, 2020c). Simultaneously, the use of nuclear power in terms of hours of utilization in China was high, at 7,394 h, in 2019, compared to 4,293 h for thermal plants and merely 2,082 h for wind in the same year (CGN, 2020a). As expected, most of CGNs revenue (61.83%) was from selling electricity to the grid system in Guangdong and Guangxi, or 37.64 billion RMB (CGN, 2020a). The operating life of the reactors does vary. For units built during 2005–2009, the lifespan is expected to be about 60 years (CGN, 2020a). The amount of electricity provided by nuclear plants increased significantly, by 25.44%, in 2019 compared to 2018. It further went up by 19.36% in the first half of 2020, even though overall electricity usage in Guangdong fell by 2.1% (CGN, 2020b, 2020a). Wholesale prices from Guangdong's plants also remained constant per kWh during 2019–2020 (CGN, 2020a, 2020b).

2.2. *Hydropower and energy storage*

Hydropower production in Hong Kong, Macau, and Guangdong has limited production potential, with Guangdong having the most installed hydro capacity. Hong Kong's and Macau's hydro supplies are severely limited. However, the reservoirs play a crucial role in providing clean water to Macau and replenishing freshwater within the GBA (The Ministry of Water Resources., 2016). The provinces around Guangdong and, by extension, Hong Kong and Macau are rich in hydropower resources, with inter-provincial hydropower trade accounting for 24.53% of Guangdong's electricity consumption in 2015 (Weigang *et al.*,

2020). In recent years, Guangdong Province has made much higher hydropower purchases. China Southern Power Grid (CSG) estimated a 99% utilization rate for Yunnan Hydro in 2020, with Guangdong averaging 200 TWh of imported power in 2019 and 2020 each (CSG, 2020a). CSG, the state-owned company that operates the power networks in Guangdong, Guanxi, Yunnan, Guizhou, and Hainan, derives 37.04% of its electricity from hydro as of 2018. Combined with nuclear and renewable energy, nearly half of Guangdong's electricity came from clean sources (CSG, 2019). These figures do not include the expected increase in import potential from the newly constructed Wudongde Dam (18.6 billion USD and Baihetan projects (both in Yuannan Province) with potential yearly outputs of up to 38.9 TWh and 62.4 TWh of electricity respectively. The Wudongde project began operations in summer of 2021 (Power Technology 2021, Proctor 2022). While the Batinertan project (24.8 billion USD) completed construction in late 2022 (Reuters 2022b) (Power Technology, 2021). However, the seasonal nature of hydro causes hydropower stations to release water without generation in wet seasons (Cheng *et al.*, 2018). Input into the system during the wet season is, on average, three times greater than that during the dry season (Liu *et al.*, 2018). This supply requires significant transmission capacity, which has not kept up with hydro production (Liu *et al.*, 2018). The extreme changes in transmission are illustrated by the fact that generation was only 1.79 TWh in February 2015 but reached 8.7 TWh in August 2015 (Cheng *et al.*, 2018). The seasonal and annual variations in the output meant that Yunnan could not provide reliable electricity supply during dry seasons or drought years (Cheng *et al.*, 2018).

Energy storage systems have been seen as one way to reduce the cost of transmission and curtailment within energy systems in the EU and China (Brown *et al.*, 2018). Hydro, battery, and hydrogen storage facilities must reduce the curtailment of renewables and moderate system costs in any low-carbon scenario (Brown *et al.*, 2018). Current battery storage with 6-h charge cycles work well with solar and wind energy. Local storage is more economical than transmission costs (Brown *et al.*, 2018). Battery technology has improved rapidly over the previous five years. The Tesla big battery was able to store only 129 MWh in 2017, while in 2020, a 230–250 MWh battery had been commissioned in California (Vorrath and Parkinson, 2020).

Further large-scale batteries are being planned with 400 or 1600 MWh of charge and eventually up to 1.5 or 6 GWh (Vorrath and Parkinson, 2020). The largest battery being built has a 1.2 GW capacity at a cost of 2.4 billion AUD (Morton, 2021). According to IRENA, energy storage needs to increase over 20 times between 2019 and 2030 to meet the Paris Climate Agreement goals, while EV storage needs to grow over 25 times (IRENA, 2020). Energy storage can avoid the cost associated with network expansion and dispatchable generators while increasing power plant utilization (Parzen *et al.*, 2021). Increasing the design freedom of energy storage can reduce the total costs of an EU grid system by up to 10% (Parzen *et al.*, 2021).

Guangdong has pumped storage hydropower capacity. However, since pumped storage uses more energy than it produces, its purpose is to generate power during peak demand. Like other energy storage facilities, it is not a practical power source by itself. In theory, pumped storage could fill the gaps in capacity if used in connection with intermittent wind and solar, which allows for the filling of the dams in pumping stations during lulls in demand and releasing the water during peak times when electricity prices are high. Guangdong is still investing in pumped storage, with 9.68 GW of capacity in operation or construction. CLP has the right to use 600 MW from the Guangzhou Pumped Power Station to make up for any gaps in generating capacity (CLP, 2021). The installed capacity is close to Guangdong's 11.37 GW theoretical maximum capacity (Guangdong Statistics, 2020). Hydropower in both pumped storage and traditional hydropower is necessary within the CSG and China to compensate for significant fluctuations in power demand due to weather events and seasonal variations (Brown *et al.*, 2018). Zhejiang, for example, saw a peak of 187.2 GWh/h in utilization on July 25, 2016, due to an extreme three-day heatwave, with this demand being 55.6% times higher than the annual average and 85.7% higher than power usage in April (Brown *et al.*, 2018). This increase in use is primarily correlated with increased air conditioning use throughout the summer, with southern China heavily reliant on cooling (Brown *et al.*, 2018).

2.3. *Waste to energy*

Hong Kong is investing in waste-to-energy (WTE) plants that may provide up to 1.5% of Hong Kong's electricity needs by 2030 (HKGOV, 2017). Hong Kong's investment in WTE is likely due to WTE's ability to reduce the harmful components in municipal waste since most of the foul gases and bacteria decompose when burned at 850–1100°C (Zhang *et al.*, 2015). Waste also accounted for 7% of Hong Kong's GHG emissions in 2019, with over 90% from landfill emissions (HKGOV, 2021a). The new (February 2021) Waste Blueprint Plan for Hong Kong 2035 shows the current capacity and output of each currently operating and expected plant (HKGOV, 2021b): (1) T-Park: opened in 2015, T-Park treats 2000 tons of sewage a day, for a total of 2 million tons since opening. The plant produces enough electricity for around 4,000 households (17.5 GWh/year). (2) O-Park: opened in July 2018, the plant converts 200 tons of food waste daily into electricity for 3,000 households. Total usage has been around 85,000 tons of food waste (14 GWh/year). (3) O-Park 2: currently being constructed for opening in 2023, the plant seeks to convert 300 tons of food waste a day into electricity (24 GWh/year), enough for 5,000 households. (4) I-Park: by far the largest of the WTE plants, with an expected operation date of 2025, I-Park is projected to handle 3,000 tons of municipal solid waste (MSW) daily. Currently, Hong Kong's MSW is 15,000 tons per day, of which 4,200 tons are waste paper, 2500 are plastics, and 3500 are food waste. The expected output (480 GWh/year) would be enough for 100,000 households.

In Macau, space is even more limited. Locally produced renewable energy is primarily from WTE plants, which provide 3.4% of Macau's electricity needs (CEM, 2020a). China has the largest solar and wind capacity globally, with over 200 GW of solar and over 230 GW of wind capacity as of 2019. It also has the highest annual investment in renewables (REN 21, 2020). As a whole, China has a high renewable potential, which may translate into a significant share of renewables within the Guangdong power mix.

Guangdong invests heavily in WTE power plants in its energy-from-waste program (EfW). China had 7.3 GW of installed capacity (339 plants) by the end of 2017 and an annual growth rate in EfW of 26% between 2012 and 2017 (IEA, 2019). Expected growth is high, with 10 GW installed by 2020 and 13 GW by 2023 (IEA, 2019). In part, WTE has become more viable due to the large amounts of MSW being produced in the country (172.39 million tons) in 2013, increasing at an annual rate of 7–8% (Zhang *et al.*, 2015). Predictions are that EfW could manage up to 260 million tons of this waste by 2025 (IEA, 2019). Shenzhen, by 2022, was planning to open a new 5000–5600 ton a day 165 MW (550 million kWh per year) plant, combusting a third of Shenzhen's daily waste output (15,000 tons) (World Economic Forum, 2019; NS Energy, 2020a). Studies conducted in Hong Kong suggest that, of all the regions in China, Guangdong had and would have the most significant potential for WTE at 10.77 TWh in 2016 and 11.82 TWh by 2025 (Wang *et al.*, 2019).

Unfortunately, emissions savings from WTE are unclear and unlikely to be viable as an effective low-carbon fuel source (CEM, 2020a). Macau's data show that its WTE plant provided only 3% of electricity in 2019. Data for WTE plants show high CO_2 emission levels per kWh, with data from 2005 suggesting an emission intensity that is 3.1 times higher than that of natural gas (Guy, 2005). A study in the UK states that a new electric-only, 265,000 tons per annum WTE plant would, on average, produce 26,447–80,454 tons of CO_2 equivalent more than that produced leaving the waste as is within a landfill (Josh, 2019). Although a WTE plant would require much less space than a landfill, more research should be conducted in follow-up studies to determine the actual carbon mitigation for the proposed WTE plants in Guangdong and Hong Kong.

2.4. *Wind and solar*

Wind and solar energy have limited potential in Hong Kong and Macau (HKGOV, 2017). Government studies show current solar and wind have a minimal presence in Hong Kong, with most Hong Kong wind power produced by a single 800 kW turbine in Lamma, while the feed-in-tariff policy facilitated the installation of about 345 MW of solar PV by 2022, with about 97.5% in CLP's electric system (HKGOV, 2017; CLP, 2021; HKEI, 2021). By 2030, Hong Kong may satisfy 3–4% of its electricity needs from local renewables, of which 1.5% is WTE and the rest is wind and solar. A small solar farm providing 1% of electricity would require 3.6 km^2 (HKGOV, 2017). A 200 MW offshore wind farm, although still in the preliminary stages, has recently gained some traction, with CLP's CEO stating that offshore wind could be economically feasible by February 2021 (RTHK, 2021). Further, CLP believes that each offshore turbine could have a capacity of 12 MW, up from 8.8 MW in 2018 (CLP, 2019). However, under new government plans, renewables are hoped to provide up to 10% of Hong Kong's annual electricity generation by 2035, divided into 4% wind, 4% WTE, and 2% solar (HKGOV, 2021a). So far, the government has invested HK$3 billion in small-scale renewable systems since 2017/2018.

Simultaneously, CSG has a much lower share, 4.15%, of its grid powered by non-hydro renewables compared to China's 7.16% (CSG, 2019). China has plans to increase renewables' share to make up 35% of all power generation capacity by 2030 (Lin *et al.*, 2020). Guangdong does have some renewable potential, with an installed capacity of 3.35 GW of wind power and 3.32 GW of solar (Liu *et al.*, 2020). Due to weather variability, solar and wind power outputs vary considerably, with wind generation ranging from 0 MW/h under no wind to 3.17 GW on a windy day (Liu *et al.*, 2020). Further, as wind conditions change throughout the day, even on moderately windy days, the average output does vary, with the typical production on a relatively windy day ranging from 573 MW to 2687 MW (Liu *et al.*, 2020). Solar similarly has issues with power output, with significant variations between seasons. In summer, power can often be generated from 5 a.m. to 8 p.m., with a maximum output of up to 1.64 GW. However, in winter, the output is only available from 7 a.m. to 6 p.m., with maximum output much lower at 637 MW (Liu *et al.*, 2020). In Guangdong, PV capacity has continued to increase from 3.32 GW in 2017 to 5.6 GW in Q2 2019 and 6.97 GW in Q2 2020 (NAE, 2019; China Energy, 2020).

However, new information suggests that Guangdong may soon increase its solar usage to 28.2 GW through some form of operation or development (GEM, 2023). This is partially expected Guangdong being one of only two provinces expected to spend more of their COVID-19 stimulus on low-carbon technologies than on fossil fuels (Myllyvirta and Yedan, 2020). Over the past few years, China has been the largest renewable energy investor

(REN 21, 2020). China added 30.1 GW of solar in 2019, nearly as much as the rest of the world combined (31.1 GW), and 26.1 GW of wind, 19.3 GW more than the rest combined (REN 21, 2020). This trend continued in 2020: over 161 billion RMB investments were in new wind energy systems by Q3 of 2020, almost 140% higher than the previous year (China Electricity Council, 2020; China Energy Portal, 2020). This led to a 10% year-on-year growth in the amount of energy produced by wind power. Total wind electricity generation in China increased 21.6 times between 2008 and 2017 (Zhu *et al.*, 2020). Solar has seen even stronger growth (National Bureau of Statistics, 2020b; REN 21, 2020).

In total, data show that China added 71.67 GW of new wind power in 2020, over double the amount in 2019. Solar has rebounded after two years of falling, with the latest solar capacity being 48.2 GW, 8.2 GW higher than what the industry estimated (Xu and Stanway, 2021). At year-end, total installed capacity stood at 281.5 GW of wind generation and 253.4 GW of solar generation capacity. This investment has led to large amounts of innovation: among the 13,160 patents filed in renewable energy, over 7,540 were from China (IRENA, 2020). Total renewable investment in China in 2020 stood at 818 billion RMB, or 30% of global investment (SCIO, 2020). Solar pricing per MWh has fallen drastically globally by 82% between 2010 and 2019 (IRENA, 2020). This pricing is expected to decrease, reaching a 58% reduction by 2030 over 2018 prices (IRENA, 2020). However, an important consideration is that offshore wind is still seen as significantly more expensive than onshore wind in China. The average cost per MWh for offshore wind in China was US$115 in 2018, compared to US$50 for onshore wind (Lin *et al.*, 2020). Total investment has increased over the COVID-19 pandemic, with China likely to significantly over-achieve their initial 2015 NDC goals of 1,200 GW of solar and wind capacity by 2030 (Climate Action Tracker, 2023). Current estimates place China's total renewable production capacity at over 1,150 GW as of 2022 (Climate Action Tracker, 2023). The total investment in 2022 for energy transition (EVs, renewables, nuclear, CCS, hydrogen, etc.) will reach US$546 billion, or nearly half the global total of US$1.1 trillion (Catsaros, 2023). With this level of investment and according to China's 14th Five-Year Plan, China should reach the goals of having 20% of energy consumption and up to 39% of electricity generation (33% renewables) in 2022 from non-fossil fuels (Climate Action Tracker, 2023).

However, China has serious problems with renewables curtailment, with 7% of wind electricity lost in 2018 (Meara, 2020). This was already an improvement over previous years, from 17.1% in 2016 and 11.9% in 2017 (Lin *et al.*, 2020). In 2017, the curtailment of wind and solar produced in windy, sunny Xinjiang and Gansu was 30% due to the inability to transmit the electricity to where it was needed (Meara, 2020). With China's high renewable energy potential in the northwestern regions, electricity transmission faces high technological and, especially, economic barriers to satisfy the electricity demand in the energy-hungry eastern provinces (Brown *et al.*, 2018).

Wind power in Guangdong, as of 2020, was at 3.26 GW from 2,567 wind turbines, excluding offshore wind farms within Guangdong waters (The Wind Power, 2020a, 2020b). Guangdong's solar potential is relatively limited, with a much lower solar energy density than western China. To compensate, Guangdong has begun investing in offshore wind farms, adding 350 MW of offshore wind in 2019. This shows a substantial increase, as the installed offshore wind capacity in 2017 was less than 50 MW (Huang *et al.*, 2020). Offshore wind will likely expand within southern China, with Guangdong, Fujian, and Zhejiang planning to build 14 GW of offshore wind soon. Guangdong is planning up to 23 offshore wind farms with a total installed capacity of 9.85 GW (Lin *et al.*, 2020). More up-to-date information from Global Offshore Wind further shows the development of offshore wind farms in the multiple-GW range (Global Offshore Wind, 2023). The largest currently proposed site, the Guangdong Deep Water Site 2, has a proposed capacity of 7.4 GW (Global Offshore Wind, 2023). Hong Kong has also begun early planning for an offshore wind farm with a rated capacity of 250 MW off its eastern coast (Global Offshore Wind, 2023). The theoretical potential of wind from offshore Guangdong is very high due to a high average annual offshore wind speed, with the theoretical potential being as high as 100 GW (Huang *et al.*, 2020).

2.5. *Geothermal*

Geothermal uses the Earth's natural heat to provide heating and electricity. As we dig deeper and approach the mantle, the temperature rises. This temperature change can provide heat (hot steam and water) and power generation. Although common in tectonically active locations such as New Zealand and Iceland, geothermal is often used mainly for heating (satisfying 90% of heating needs in Iceland). As an energy source, geothermal power provides several advantages over other power sources, including much lower land requirements and emissions and higher baseloads (IEA, 2020). Geothermal electric generation facilities are located primarily in the United States, with 2.59 GW of capacity in 2019, followed by Indonesia with 2.13 GW and the Philippines with 1.9 GW, while China has about 400 MW of capacity (IRENA, 2019). If accounting for heating, such as ground-source heat pumps (55.3%) and water heating for pools (20.3%), the global geothermal capacity would increase to 70.3 GW of heating capacity in 2014 (Lund and Boyd, 2015). China's capacity would increase as well to 17.9 GW, of which 11.8 GW is in heat pumps, with geothermal contracted in

Shenyang (22.48%) and Beijing (15.18%) (Lund and Boyd, 2015). This concertation of using geothermal for heating is apparent in most countries, with the United States' ratio of heat pumps to power generation capacity at 5.12/1, or 19.7 GW vs. 3.85 GW. France has 16.7 MW of electric capacity compared to 2.6 GW of direct-use capacity (including 2 GW heat pumps) (IEA, 2020).

Iceland makes the most use of geothermal electricity, responsible for about 30% of the national generation (IEA, 2020). For Guangdong, the first large-scale geothermal project was in 1970, with a plant in Fengshun, although this plant has been discontinued. High-temperature geothermal resources are located close to tectonic plate boundaries near the Himalayan mountains, far away from Guangdong. Guangdong and other coastal provinces have access to low- and medium-temperature geothermal belts with over 6,700 km^2 of area that could be developed for geothermal exploration (Zhang, Chen and Zhang, 2019). Other papers recommend that Guangdong invest in low-temperature pilot programs to exploit geothermal energy (Zhang *et al.*, 2019).

3. Electricity Transmission

3.1. *Intra-regional*

Hong Kong has no ultrahigh voltage (UHV; >800 kV), and its electricity is managed by two vertical monopolies with separate service territories, namely CLP and Hong Kong Electric (HKGOV, 2018). CLP's transmission is at the 400 kV level, with a total of 33.2 TWh in 2017, while Hong Kong Electric's transmission is through lines at the 275 kV level, with a total of 10.6 TWh in 2017 (HKGOV, 2018). In 2019, Hong Kong had a peak demand of 9.6 GW and an installed capacity of 12.23 GW (HKGOV, 2020). Due to the short distances in Hong Kong, HK Electric had a transmission loss rate of 3.4% in 2019, whereas CLP's corresponding rate was 3.83% on average between 2015 and 2019

(CLP, 2020; HK Electric, 2020). CLP controls about 16,135 km of transmission & distribution lines at or above 11 kV in 2020, including 555 km at 400 kV, 1,671 km at 132 kV, 22 km at 33 kV, and 13,782 km at 11 kV (CLP, 2020a), while the fuel mix of electricity generation in 2019 was 29% natural gas, 36% coal, 35% nuclear, and less than 0.5% others (CLP, 2020b). CLP also received electricity from the Daya Bay Nuclear Plant in Shenzhen and the Guangzhou Pumped Storage Hydropower Plant in Guangzhou via dedicated transmission lines to Hong Kong. Greater grid integration could decarbonize Hong Kong's electricity consumption due to the higher shares of non-fossil electricity with China Southern Power Grid (CSG, 2020b). Before 2018, Hong Kong regularly exported fossil-fuel-fired electricity to Mainland China; however, the quantity decreased significantly with no exports since 2019 (National Bureau of Statistics, 2020a). The total asset value of the transmission & distribution system in 2019 was a little under HK$80 billion (CLP, 2020a). The total transmission system accounts for 48% of CLP's total capital assets, or HK$92.54 billion in value, while Hong Kong's share is just under HK$80 billion (CLP, 2020a). Hong Kong Electric had a transmission & distribution network of 6,540 km and a total generation power of 3,237 MW (HKEI 2020), while information on the line length at each voltage level is unavailable (HK Electric, 2020).

Macau's electricity transmission is handled by CEM, which holds 1,010 km of high-voltage lines rated between 66 and 220 kV. Lower-level transmission lines between 400 V and 11 kV extend for 4,005 km, primarily via underground cables (CEM, 2021). Peak demand in Macau during 2020 was 955.5 MW, with 93% coming from the CSG grid (CEM, 2021).

CSG provides power to five provinces: Guangdong, Yunnan, Guangxi, Guizhou, and Hainan, as well as Hong Kong and Macau. It distributed electricity in Guangdong via three main branches: the Guangdong Power Grid (31.5 million customers), the Shenzhen Power Supply Bureau (2.88 million customers), and the Guangzhou Power Supply Bureau (5.45 million customers) (CSG, 2015). Guangdong Power Grid is a CSG wholly-owned subsidiary in charge of 19 city power supply bureaus. The system is centered around the Pearl River Delta with a 500 kV network of eight AC and eight DC lines. It is also connected to Hainan (one 500 kV line), Hong Kong (four 400 kV lines), and Macau (six 220 kV lines and three 110 kV lines). Data on total line length are unavailable, while CSG operates 237,000 km of high-voltage lines (110 kV and above) across the five provinces. Shenzhen Power Supply provides power to all of Shenzhen except Shekou. In 2016, it supplied 81.444 TWh and had a maximum capacity of 16.26 GW. Guangzhou Power Supply Bureau provides power to 11 districts in Guangzhou. In 2016, its 110 kV and above transmission line length was 7,181 km.

3.2. *Inter-regional*

China has two transmission grid corporations, namely the CSG and the China State Grid, that control all other provinces. China's grid system is extensive, with over 540,000 km of power lines rated at 220 kV or above at the end of 2012 (Wei, 2016). This has increased rapidly, with China State Grid alone owning 1.09 million km of transmission lines (110 KV and above) at the end of 2019 (State Grid, 2019). At the same time, CSG controls 237,000 km of power lines (110 kV and above), with nearly half, or 102,141 km, rated above 110 kV (CSG, 2019). Each grid system primarily operates independently, with limited trade between grid systems and a transmission capacity of 179 GW in 2017 (Xu *et al.*, 2020). China has begun investing heavily in 800 kV or above UHV lines. Between 2011 and 2015, China State Grid alone invested US$43 billion in building 40,000 km of UHV lines (Paulson Institute., 2015). China had 21 UHV lines completed and signed off 12 new lines by 2019. Additional construction was on-going with 11 projects being constructed in June

2022 under State Grid with an additional 22 billion USD being earmarked for the second half of the year (Reuters, 2022a). Within 2024 total investment into Chinas grid system is expected to reach 500 billion yuan nearly 70 billion USD with six UHV alternating currents (1000 kv +) to be completed during the year (Zheng, 2024). In addition State Grid predicts that under the 14th Five Year Plan (2021–2025) total investments in UHV will reach 380 billion yuan 35.8% more than the previous plan (Zheng, 2024). Total UHV projects likely to be completed under the plan is expected to reach 38 (Zheng, 2024). The availability of these high-voltage transmission lines in both grid systems might partially explain the reduction in transmission loss, which was reduced from 6.83% in 2013 to 6.25% in 2019 in China State Grid and from 6.72% in 2015 to 6.31% in 2018 in China Southern Power Grid (CSG, 2019).

The building of UHV lines in China is rationalized by the unmatched geographical distributions of hydro, coal, and renewables in western and northern China, on the electricity generation side, and population and economic activities in eastern and southeastern China, on the electricity consumption side (Paulson Institute, 2015). If counting high-voltage lines of 500 kV or above, China Southern Power Grid operates 19,470 km of transmission lines between its five provinces (CSG, 2019). One reason for increasing the high-voltage lines is to reduce the curtailment of renewables in western provinces (Meara, 2020). Five high-voltage DC (HVDC) lines can carry 26 GW of hydropower from Yunnan to Guangdong, meeting a quarter of Guangdong's electricity demand (Fairley, 2016). CSG plans to invest 120 billion RMB in new infrastructure by 2023, with a new HVDC line connecting the GBA with the Wudongde Hydropower Station (CSG, 2020a).

A further transmission line infrastructure in the form of five AC and two DC lines is between Guizhou and Guangdong Provinces, with a transmission potential of 11 GW (CSG, 2015). The 13th Five-Year Plan hoped to increase this amount with 19 lines across all five provinces (8 AC and 11 DC) to allow for a transmission capacity of 58 GW (CSG, 2020b). However, the current estimate is 46 GW capacity (Cheng *et al.*, 2018). Imported electricity from western China meets about a third (35 GW) of Guangdong's consumption, a majority of which, 174.7 TWh, or 84.7%, is non-fossil-fuel fired (CSG, 2020b; Zhu *et al.*, 2020).

In 2015, the installed generation capacity of CSG was 243 GW, of which 42.1% (102.3 GW) was hydropower, while its peak load was only 141 GW (Zhou *et al.*, 2016). CSG's total capacity as of 2019 was expected to be 320 GW (CSG, 2020b). In 2018, CSG sold 970.3 TWh of electricity through the five provinces, up from 782.2 TWh, or by 24%, since 2015, while total assets also increased by 28%, from 636.2 billion RMB to 816.7 billion RMB between 2015 and 2018 (CSG, 2019). Total electricity generation within Guangdong in 2019 was 343.4 TWh, up from 333.3 TWh in 2017, while total consumption was 669.6 TWh (Guangdong Yearbook, 2020). In 2020, the maximum Guangdong grid load reached 127 GW (CSG, 2020b). Transmission data show that Guangdong in 2017 received 206.5 TWh of electricity from Yunnan, Guizhou, Hubei, and Guangxi (Zhu *et al.*, 2020). The breakdown of interprovincial transmission in 2015 was 18.9 GW from Yunnan hydropower, 8.6 GW from Guizhou, and 3 GW from the Three Gorges Dam (Zhou *et al.*, 2016). Transmission capacity in 2017 was very similar, with Guangdong having 30 GW from southwestern China and 3 GW from the Three Gorges Dam (Xu *et al.*, 2020). However, some research suggests that investment in UHV lines is excessive, and unidirectional DC lines from western to eastern China are sufficient to reduce western China's renewable curtailment. However, the lack of interconnection between State Grid and CSG raises feasibility issues (Brown *et al.*, 2018). Overall, transmission capacity from western China to eastern China was estimated to be 229.1 GW in 2017, and it could increase to 540 GW by 2030 (Zhu *et al.*, 2020). Total transmission from west to east was at 682.5 TWh in 2017, and by 2030, it is estimated that 40% of western China's electricity production will be sent eastward, with annual transmissions reaching nearly 1,900 TWh (Zhu *et al.*, 2020).

4. Assets of Non-Fossil-Fuel Infrastructures

The energy transition is rapidly reshaping the energy mix of Hong Kong and the GBA and non-GBA Guangdong Provinces. In 2020, our data show that the entire region had 8,856 MW of solar and wind power, constituting 6.5% of total capacities; while counting only the current projects in the pipelines, the numbers are expected to grow to 46,992 MW and 22.3% in 2025 and 59,452 MW and 25.7% in 2030 (Figures 1 and 2). Nuclear power capacity will also jump from 16,136 MW in 2020 to 28,136 MW in 2030. Coal-fired power capacity barely increased over the decade. With the significant growth in natural gas-fired power capacity, one kWh of electricity generation in the region will be much less CO_2-intensive.

Another significant feature is that GBA and non-GBA Guandong have distinct energy transition trends. As analyzed in Chapter 2, GBA (including Hong Kong) is witnessing an energy transition from coal to natural gas, while non-GBA Guangdong is transitioning from fossil to non-fossil fuels (Figure 3 in Chapter 2). In 2020, non-GBA Guandong had 7,567 MW of wind and solar capacity, which is expected to grow to 39,902 MW in 2025 and 51,126 MW in 2030, counting only the current projects in the pipelines (Figure 3 in Chapter 2). In 2025, only 6.7% of electricity generation capacities in GBA Guangdong are expected to be wind and solar, while the ratio in non-GBA Guangdong will

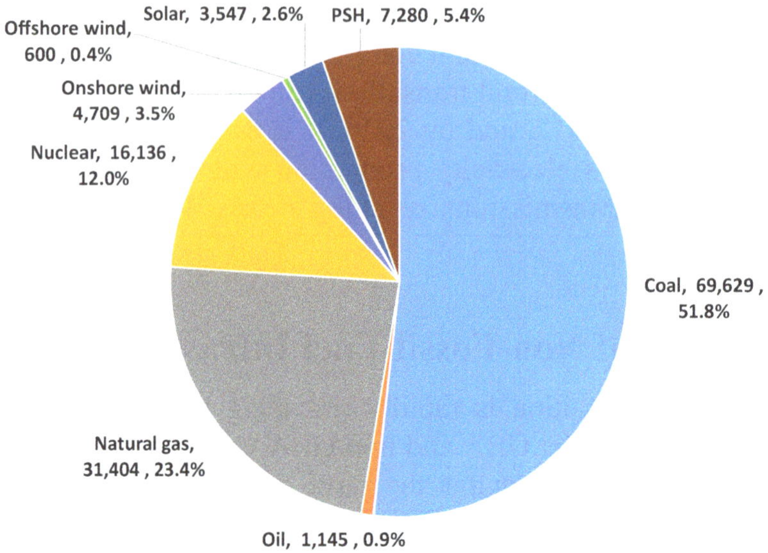

Figure 1. Electricity generation capacity in 2020 by fuels.

reach 40.3% (Figures 2 and 3 in Chapter 2). Of the wind and solar capacities in Hong Kong and Guangdong Province in 2025, 84.9% are expected to be located in non-GBA, and the ratios for onshore wind, offshore wind, and solar PV will be 87.2%, 93.6%, and 71.9%, respectively (Figures 2 and 3 in Chapter 2). The patterns remain very stable throughout the decade, being 85.4% in 2020 and 86.0% in 2030, with Zhanjiang, Yangjiang, Qingyuan, Shantou, and Shanwei taking the lion's share of the total due to their much lower populations and GDP densities (Figures 3 and 4).

Solar PV and offshore wind are expected to grow especially strongly, from 3,547 MW and 600 MW in 2020 to 17,044 MW and 23,764 MW in 2025 and 28,504 MW and 24,764 MW in 2030, respectively (Figures 1 and 2). The slower increasing pace from 2025 to 2030 is because only those existing and planned

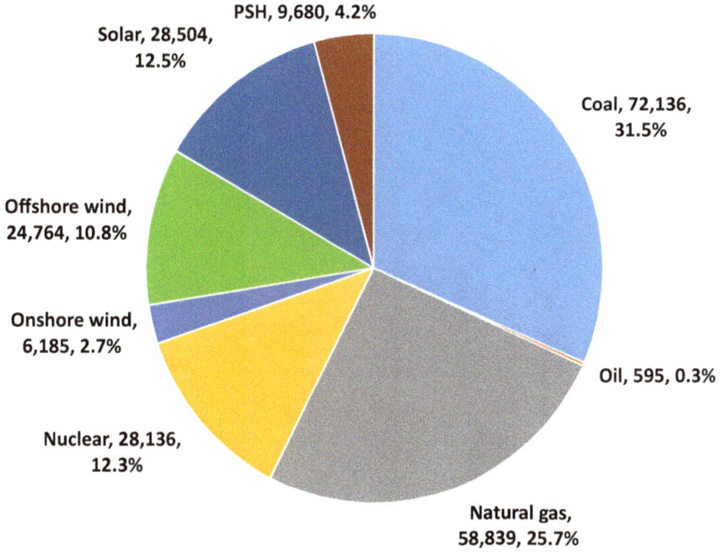

Figure 2. Electricity generation capacity in 2030 by fuels (including those planned and retiring power plants).

pipeline projects are counted here. In contrast, onshore wind development is relatively limited and set to rise slightly from 4,709 MW in 2020 to 6,185 MW in 2030.

Guangdong is also adding a significant number of nuclear power plants, and the capacity is expected to increase from 16,136 MW in 2020 to 28,136 MW in 2030. They are more evenly distributed between GBA and non-GBA Guangdong when those projects in the pipeline are completed. In 2020, Shenzhen and Jiangmen in the GBA had 6,120 MW and 3,500 MW of nuclear power capacities, respectively, and Yangjiang in the non-GBA had 6,516 MW. In 2030, Huizhou (GBA) and Shenwei and Zhanjiang (non-GBA) are also expected to have nuclear power plants.

From the perspective of asset values, Hong Kong's non-fossil assets are insignificant. For the entire region of Hong Kong,

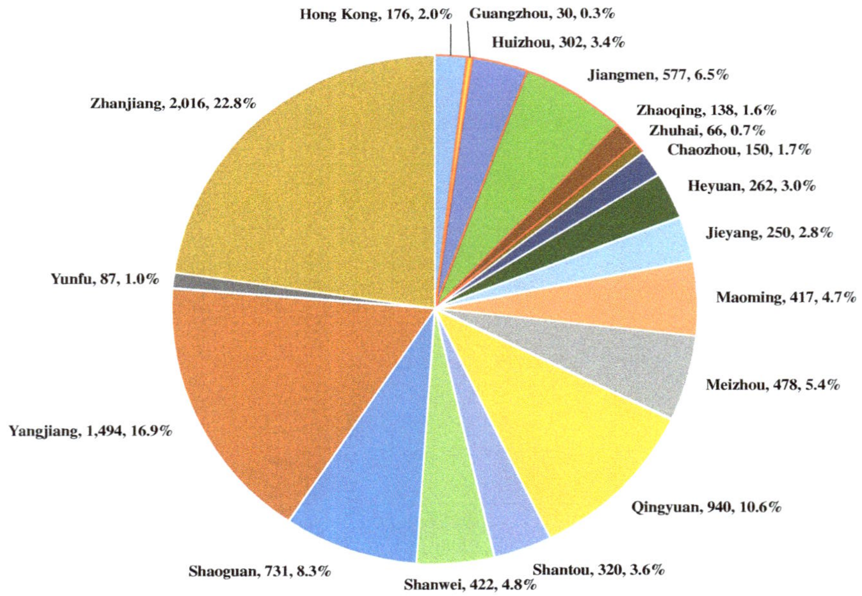

Figure 3. Renewable electricity generation capacity (MW) in 2020 by municipalities (solar PV, onshore wind, and offshore wind).

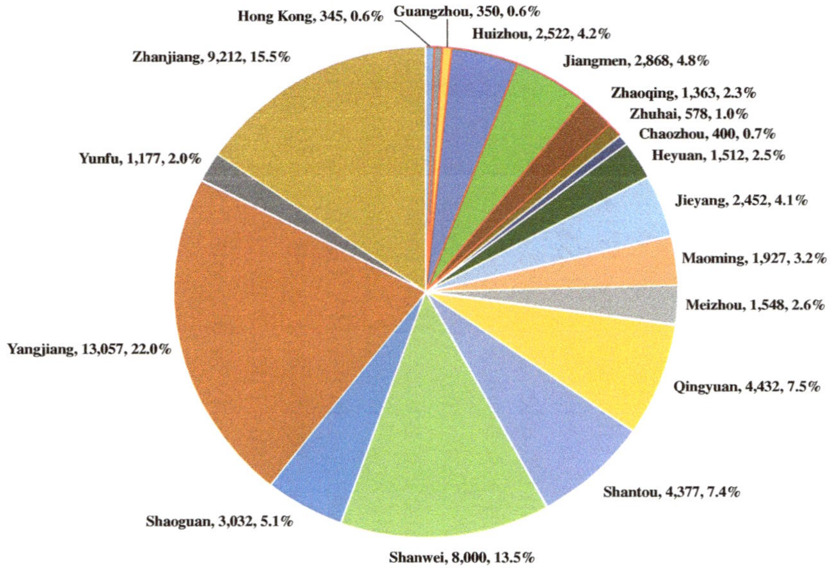

Figure 4. Renewable electricity generation capacity (MW) in 2030 by municipalities (solar PV, onshore wind, and offshore wind; including those planned and retiring power plants).

Macau, and Guangdong Province, in 2020, the asset values of wind and solar power plants were US$5.1 billion, whereas those of nuclear and pumped storage hydropower (PSH) capacities were US$30.6 billion. Fossil-fuel assets, including power plants, LNG terminals, and oil refineries, were worth US$82.8 billion. Counting retirement, depreciation, and new investment, the asset values in these three categories in 2030 are expected to be US$33.3 billion, US$53.6 billion, and US$85.7 billion, respectively. According to the project timelines, the region's non-fossil-fuel assets will surpass fossil-fuel assets by 2030.

The asset values of renewables are concentrated in non-GBA Guangdong. In 2025, GBA and non-GBA Guangdong will have asset values of US$3.6 billion and US$30.7 billion, respectively (Figures 6 and 7 in Chapter 2). These existing and planned projects will have their asset values depreciated to US$0.2 billion and US$3.4 billion in 2050, respectively, and to negligible levels in 2060. Existing nuclear power plants today and those in the project pipelines will dominate the depreciated assets in 2050 or 2060 in both GBA and non-GBA Guangdong. Like fossil-fuel assets, the remaining renewable assets in 2050 or 2060 will not be significant, while only nuclear assets will be subject to future policies on their probability of becoming stranded assets.

Chapter 4

Energy Infrastructure Gap of Carbon Neutrality

1. Electrification Gap of Energy Services

1.1. *Hong Kong's electrification*

In Hong Kong, 57% of end-use energy consumption in 2019 was electricity (electricity consumption is converted based on its thermal content without adjustment using electricity generation efficiency) (Electrical and Mechanical Services Department, 2021). By breaking down the energy services, we can analyze current electrification status in Hong Kong in great detail. We allocate CO_2 emissions from electricity generation to those services that utilize electricity, which then become those energy services' Scope 2 emissions (Figures 1 and 2). Most energy services in the industrial, residential, and commercial sectors that are associated with CO_2 emissions are fully electrified, with values ranging from just over 70% electrified in the industrial sector to over 90% electrified in the residential and commercial sectors (Figures 1 and 3). Electrification is an essential step for decarbonizing energy services with non-fossil fuels, especially renewables that should be largely converted into electricity before consumption.

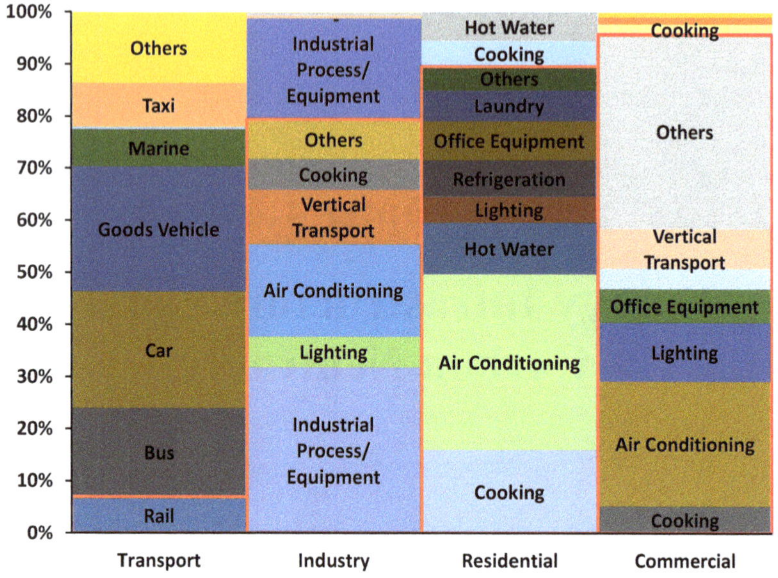

Figure 1. The electrification rates of *Hong Kong's CO_2 emissions by sector and energy services in 2019* (estimated by the authors using Hong Kong energy end-use data from Electrical and Mechanical Services Department, 2021; red rectangles indicate CO_2 emissions from electricity).

A great majority of vehicles in Hong Kong are still using oil products as their primary fuel source (Figure 3). Although the transport sector is not the highest emitting sector within Hong Kong, it is by far the least electrified and responsible for most of the non-electricity CO_2 emissions (Figures 2 and 4). Although challenging, electrifying the transport sector is possible, with Shenzhen offering a compelling example of a nearly fully electrified public transportation market, with all taxis and public buses fully electrified (Shen, 2022). Electric vehicles (EVs) and boats are vital to electrifying a lion's share of unelectrified CO_2 emissions. Current technologies can electrify almost all passenger vehicles, buses, light-duty vehicles, and small boats.

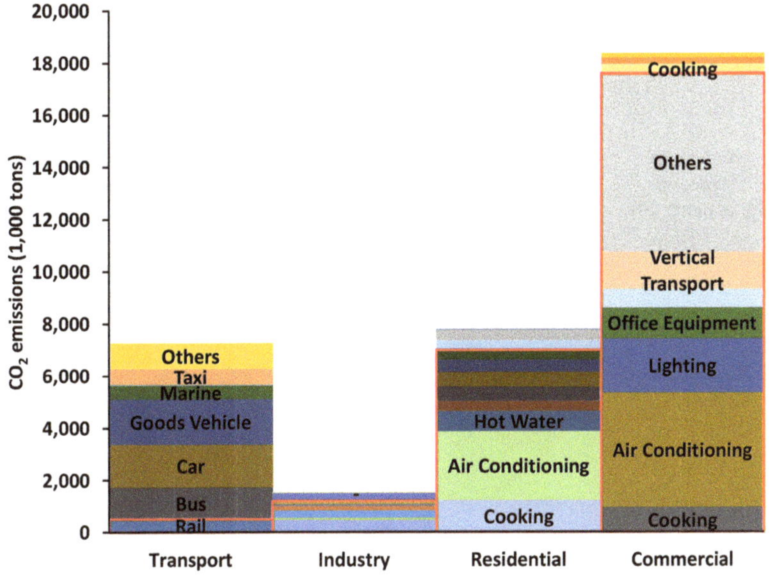

Figure 2. Hong Kong's CO_2 emissions by sector and energy services in 2019.

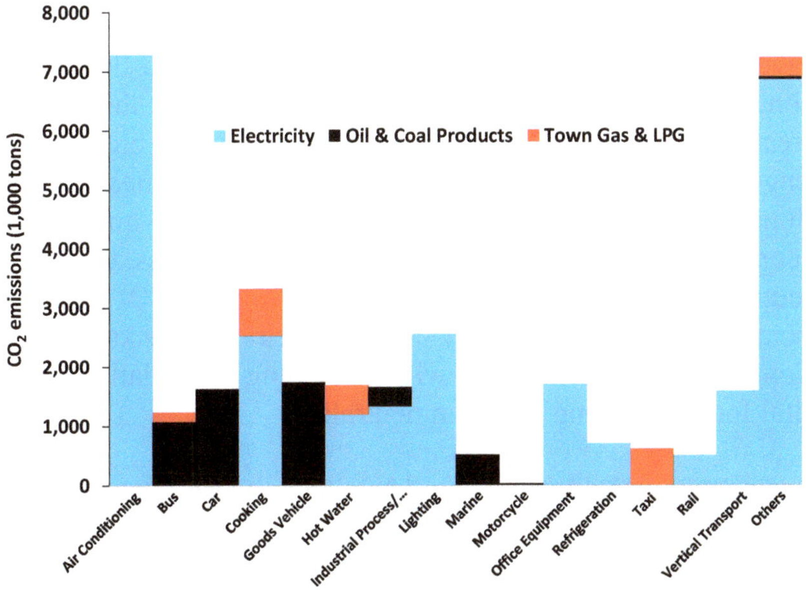

Figure 3. Hong Kong's CO_2 emissions by energy services and sources in 2019.

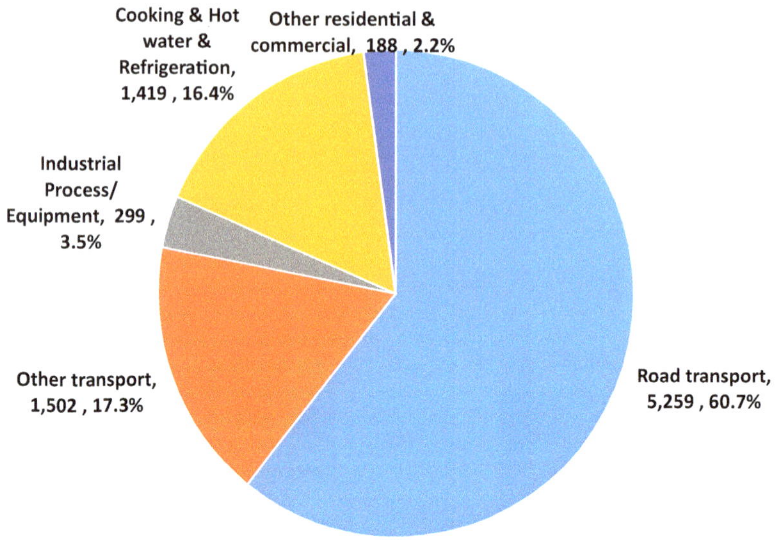

Figure 4. CO_2 emissions from non-electrified end use in 2019 in Hong Kong (unit 1000 tons).

Hong Kong also faces additional problems that may limit the adoption of EVs, with the hilly topography requiring EVs to have large batteries for similar ranges compared to flatter locations. At the same time, higher humidity increases the need for greater maintenance of EVs (Tsoi *et al.*, 2022). The biggest challenges are faced by ocean-going ships and airplanes. Because Hong Kong's official CO_2 emissions inventory only counts Scope 1 emissions within its geographic boundaries, the carbon neutrality pledge includes only a tiny fraction of CO_2 emissions from shipping and aviation that fill their fuel tanks in Hong Kong. If the scope becomes wider, shipping and aviation will be vital for Hong Kong's carbon neutrality.

Other areas, such as cooking and heating, can also have significant impacts. Induction cookers could electrify cooking and replace town gas and LPG, while electric water heaters are commercialized, mature applications that could replace current gas water heaters. Electric motors can also replace most diesel engines in industrial processes/equipment. As a result, the electrification of Hong Kong's unelectrified CO_2 emissions is mainly affected by economic calculations but not technological challenges.

1.2. *Electrification in Guangdong*

Energy services in Guangdong Province are much less electrified than those in Hong Kong (Figures 5 and 6), primarily because non-GBA Guangdong is less developed. Increased electric usage within Guangdong could have direct carbon emission reduction benefits in line with the associated emissions factors (EIA, 2023). Progress has been made in the electrification of transport through EV adoption within the province and greater electrification in manufacturing (Figure 5) (Global Energy Monitor,

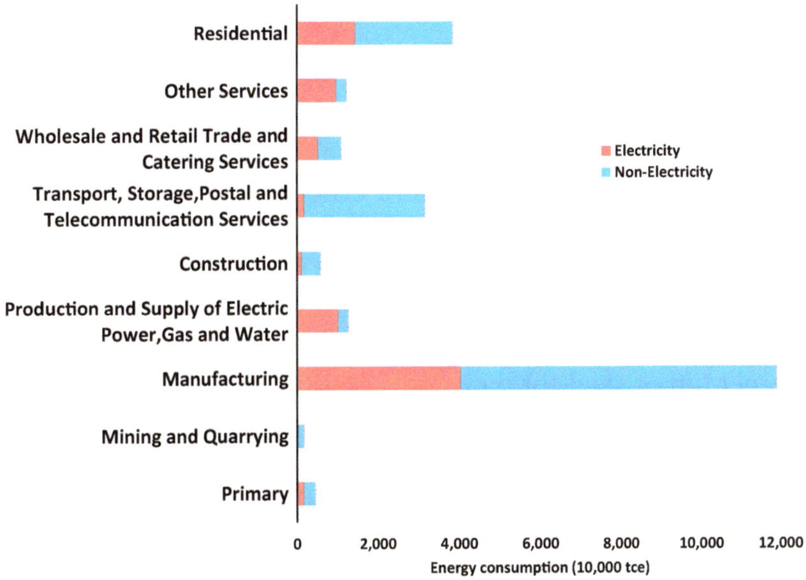

Figure 5. Electrification status of energy consumption in Guangdong (unit tce: tons of coal equivalent; electricity consumption is converted based on its thermal content without adjustment using electricity generation efficiency) (Guangdong Bureau of Statistics and National Bureau of Statistics, 2022).

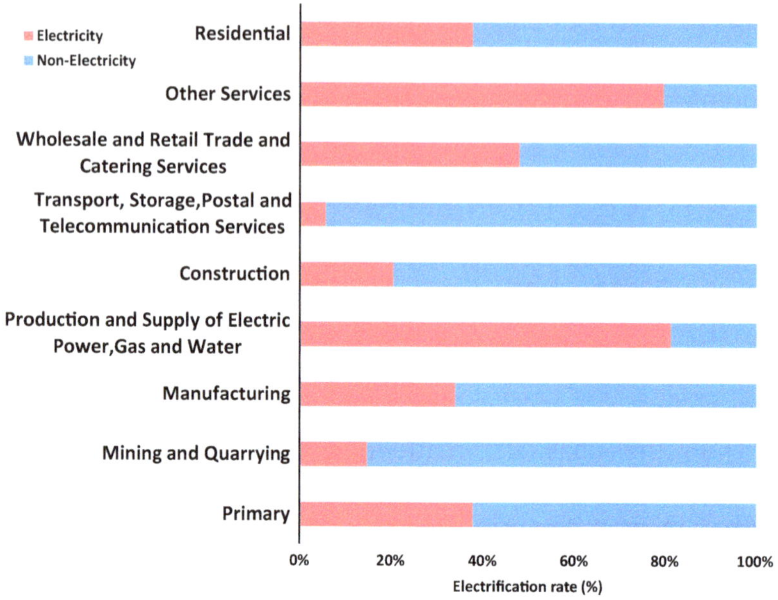

Figure 6. Electrification rates of energy consumption in Guangdong (Guangdong Bureau of Statistics and National Bureau of Statistics, 2022).

2023). The electrification of steelmaking is enabled primarily through the introduction of more efficient electric arc furnaces (EAFs) within Guangdong (Global Energy Monitor, 2023). According to BHP (2020), depending on the fuel source (natural gas) and the inputs (pre-heated scrap metal), electrification through EAFs can reduce the associated carbon emissions from 2 tons of CO_2 per ton of steel for typical blast furnaces down to 0.4 tons of CO_2 per ton of steel (BHP, 2020). However, a more likely case is that scrap is readily available for the entire production system in an EAF system; hence, iron ore will still need to be melted down to produce steel. Additionally, steelmaking in Guangdong uses over 50 million tons of metallurgical coal as a critical steelmaking component in traditional blast furnaces. Changing to electric furnaces would significantly reduce this industry's associated emissions (Guangdong Bureau of Statistics and National Bureau of Statistics, 2022).

Further growth of EVs is still expected within the greater GBA region. China accounts for most of global EV sales and over half of all EVs on the road (HKGOV LEGCO, 2021). Commercial vehicle deployment is also very advanced, with China hoping to have over 1 million commercial EV vehicles in operation by 2030 (Lumiao and Zhanhui, 2020). On a national level, the share of EV buses in China increased from 1% in 2013 to 55% in 2019, while all public buses in Shenzhen and Guangzhou were replaced with EVs (Lumiao and Zhanhui, 2020). In the GBA, current EV adoption has led to a 1.73 million ton reduction in CO_2 emissions over a base case BAU scenario without EVs (Tsoi *et al.*, 2022). However, although EVs have achieved market penetration in the GBA, to reach deep decarbonization targets by 2050 (60% below 2010 emissions), almost all new vehicles, both commercial and private, will need to be EVs by 2031 (Tsoi *et al.*, 2022).

To meet these goals, the average density of charging stations as crucial infrastructures would need to increase exponentially to support the influx of vehicles (Tsoi *et al.*, 2022). Meanwhile, the CO_2 intensity of EVs needs to be reduced by 85% through significant changes to the electric grid (Tsoi *et al.*, 2022). The need for this high level of adoption and reworking of the energy and transport sectors was expatiated by examining historical data from 2005 to 2018 on vehicle ownership in China and the GBA. If current buying trends continue, even if China reaches EV adoption of 50% by 2035, overall emissions will still grow by 54.2% (Tsoi *et al.*, 2022).

1.3. *Global trends and EVs within China*

Electrification of energy services and decarbonizing electricity are the technologically most mature pathways to carbon neutrality. At the same time, the transport sector has the lowest electricity rates in both Guangdong and Hong Kong (Figures 1 and 6). From a global perspective, the planned public EV infrastructure will likely be insufficient for the size of the targeted EV market (IEA, 2022). Further short-term geopolitical issues can significantly affect EV pricing due to the extensive requirements of materials such as cobalt, lithium, and nickel (IEA, 2022). The price of lithium significantly increased by 700% in early 2022 due to the Ukraine–Russia conflict (IEA, 2022).

Global private EV sales were projected to reach 37% of the vehicle market share by 2030 (Hsieh *et al.,* 2020). This likely rise in sales is partly due to China's current low car ownership rate compared to the United States (Hsieh *et al.,* 2020). These projections should be updated to reflect the much accelerated EV sales in the past few years. EV sales in China for August 2020 were higher than those of traditional vehicles (Narayanan, 2020). With China selling 1.3 million new electric and hybrid cars in 2018, a 62% rise over 2017, this number is even more impressive considering that, in September 2018, there were only an estimated 4 million EVs on the road worldwide (Thornton, 2019). The Chinese government announced targets to increase the EVs' market ratio to 12% of all vehicle sales by 2020, up from 4% in 2018 (Thornton, 2019). This was expected to increase to 20% by 2025 (Argus, 2020), while the actual ratio has already been over one-third in 2023.

In 2021, EV sales doubled from 2020 sales to over 6.6 million vehicles, with one in 10 cars sold being an EV. Of this, over half were sold in China, with an additional 2.3 million sold in Europe and a further 630,000 in the United States (IEA, 2022). Europe's growth meant that by the end of 2021, over 5.5 million EVs were on the road, more than three times that in 2019, while 86% of vehicles sold in Norway in 2021 were EVs (IEA, 2022). The trend continued in 2022, with YOY sales up by 71% in Q3 of 2022. China, in particular, saw 100% growth in Q3 2022 (Mukherjee, 2022). EVs had increased to 15% of all vehicles sold and become the preferred choice of vehicle for first-time car buyers in developed regions (Mukherjee, 2022).

Sales and production numbers decreased temporarily at the beginning of the COVID-19 pandemic, with China's "new energy vehicle" (NEV) production from January to September 2020 being 18.7% lower than that a year earlier and sales falling by 17.75% (Argus, 2020). Official government data show slightly different numbers, with sales falling 11.4% for NEVs between 2019 and 2020. For EV systems in the GBA, the main focus must be on Guangzhou/Shenzhen, not Hong Kong or Macau. This is due to the high public interest in buying EVs and government support (Narayanan, 2020). Guangdong province was the fourth-largest battery EV market during 2011–2017 (77,378 units). Beijing was the largest market for EVs, with 144,600 units sold between 2011 and 2017 (Zheng *et al.*, 2020). The first 11 months of 2021 saw a rebound (Shen, 2022). During the first half of the year, total EV sales were 201.5% higher than in 2020, with over 1.2 million EVs sold compared to the total car sales increase of 25.6% (BYD, 2021). This trend continued throughout 2022, with significant growth in EV sales, especially in China, with Build Your Dreams (BYD) becoming by far the largest EV seller in Q3 2022 (Mukherjee, 2022).

The Shenzhen government, in particular, is heavily sponsoring the adoption of EVs (Gray, 2018). In 2018, Shenzhen became the first major city to run a fully electric bus system, with 16,000 buses (Gray, 2018). With a further goal of turning all taxis (22,000) electric by 2020, all of which are provided by BYD, a Shenzhen-based car company (Gray, 2018). To facilitate this, Shenzhen was building 5,000 taxi-only EV charging posts to be finished by 2020 (Hove and Sandalow, 2019). Issues of capacity and utilization should be raised for public and at-home charging. Informal "fly-line" extension cord charging from older apartment blocks to EVs could cause grid reliability and distribution issues as older neighborhoods have insufficient infrastructure for large amounts of in-home charging (Hove and Sandalow, 2019). Further spikes in EV charging usage are common on weekends and public holidays, with highway charging doubling during Chinese New Year 2018 compared to the month before (Hove and Sandalow, 2019). By the end of 2019, 400,000 buses and 430,000 trucks ran on new energy sources (hydrogen, EVs, and BEVs) through China (SCIO, 2020).

For BYD and the Shenzhen government, as diesel buses accounted for roughly 20% of the city transport emissions, the introduction of electric buses led to a reduction of 48% in CO_2 emissions compared to diesel buses and a 100% reduction in other pollutants (Gray, 2018). In total, estimates were that BYD achieved a decrease of 980 million liters of petrol and reduced CO_2 emissions by 2.22 million tons (Greater Bay Insight, 2019). BYD, in 2018, was globally the second-largest manufacturer of electric passenger vehicles (233,000 units) and the largest if we include trucks and buses (247,000 units). In mid-2019, BYD commanded a Chinese EV market share of 24%. These numbers have increased significantly over the past two years. In 2021, BYD alone saw a jump in sales to over 590,000 vehicles, of which 54% were all-electric and 46% were plug-in hybrids; this is an increase of over 230% from 2020 (Shen, 2022). Further growth is expected for BYD, with sales predictions of 1.3 million EV deliveries over 2022 (Shen, 2022). BYD is also expanding its NEV public transport offerings, with new deals for NEV public transport in six new Chinese cities. The revenue associated with this is considerable. From January to June 2021, BYD recorded a revenue of over 38 billion RMB (BYD, 2021). To facilitate this influx of electric vehicles, Guangdong was building 180,000 charging points and 200 hydrogen fueling stations by the end of 2022 (Argus,

2020). China installed 17,000 EV charging stations monthly from December 2018 to November 2019, while the number further increased to 33,000 units per month (Yuanyuan, 2020). Total EV charging locations by the end of 2019 had reached 500,000, up from 50,000 in 2015 (Wu and Yang, 2020). A total of 7,400 charging piles had been built on 942 expressways nationwide (SCIO, 2020). Before the COVID-19 pandemic, the expected government investment was high, with the Chinese government planning on an additional 2.5 million charging units and 7,400 fast-charging stations in tier-one cities and provinces (Qiao and Lee, 2019). China's goal was to have 12,000 charging stations and 4.8 million charging posts by 2020 (Qiao and Lee, 2019).

The lack of charging stations has been recognized as a significant barrier to EV adoption, although studies have focused on optimizing location rather than how charging points affect sales (Wu and Yang, 2020). A study demonstrates that changing the number of available charging points by one standard deviation will affect the number of EVs on the road by 13%, while this effect is much higher on public charging points, where a one-standard-deviation change affects sales by 23% (Wu and Yang, 2020). However, there are many barriers to setting up additional charging points since more charging stations in high-traffic areas must be used due to slower charging times compared to gas-driven vehicles. Issues of acceptance for installing charging posts have been encountered with office, retail, residential, and government property management companies (Wu and Yang, 2020). Simultaneously, space issues in parking garages and the need to rewire existing electrical systems in parking structures can also become challenging problems to overcome (Wu and Yang, 2020). For Hong Kong and the GBA, which are highly developed, their energy services have been highly electrified.

In Hong Kong, at the end of October 2020, there were 16,829 EVs for road use, up from 100 at the end of 2010, while a further 3,219 EV chargers were available for public use (HK EPD, 2020). The total number of registered private cars in Hong Kong was 660,000 (HKGOV, 2021a). The number of personal EVs would increase to 21,000 in 2021, with one out of five new purchases being an EV (HKGOV, 2021a). Growth in private car EV sales did increase significantly between year-end 2021 till year end 2023 with 19,795 new registrations for EV's in 2022 (up from 9,583 in 2021) and 28,541 new registrations in 2023 (Transport Department, 2024). However, commercial EV vehicle sales and market share growth are behind Hong Kong's private vehicle sales. Nor are current and expected government policies and investments likely to change this trajectory (HKGOV, 2021b).

As of 2020, government support for EV users primarily consists of waiving tax to purchase a private EV vehicle for the first HK$97,500–250,000 spent, depending on whether the car is a replacement vehicle (HK EPD, 2020). Further, the government has allocated HK$180 million to purchase 36 electric buses and invested HK$2 billion in installing more EV charging locations (HK EPD, 2020). In its policy address in November 2020, the HK government mentioned that it would examine plans to launch a roadmap to cease the sale of conventional fuel-powered private cars (HKGOV, 2020c). The report was made public in March 2021, with financial highlights of an HK$200 million Green Tech Fund and HK$260 million to promote EV public buses (HKGOV, 2021b). Overall, the plan hopes for no new registrations of fuel-powered private cars, including hybrids, in Hong Kong by 2035 (HKGOV, 2021b). The previous maximum subsidy has also been raised from HK$250,000 to HK$287,000 under the one-to-one replacement scheme, which remains in place until March 2024 (HKGOV, 2021b). This has been the most significant policy in Hong Kong, with concessions of over HK$7.4 billion between 2015 and 2021 (HKGOV, 2021b). As for charging stations, the goals are 150,000 private charging ports by 2025 and 5,000 public spots, with an overall goal of zero vehicle emissions by 2050 (HKGOV, 2021b). In 2021, 3,351 public charging stations were available, including 1,252 of standard size, 1,456 of medium size, and 643 quick-charging stations (HKGOV, 2021b). Regarding the charging network, HK$190 million was made available in 2019 for government charging spots, a further HK$2 billion for home charging was released in October 2020, and a final HK$1.1 billion in the new energy transport fund was released for trials and applications of green transport (HKGOV, 2021b). Current emissions from vehicles made up about 16% of Hong Kong's total emissions in 2018, while that of the transport sector as a whole was 18.1% of the total (Figure 2), with the majority of non-electrified emissions relating to road transport emissions (Figure 4) (HKGOV, 2020a, 2021b). The Hong Kong SAR government understands that for Hong Kong to achieve net zero by 2050, deep cuts in emissions from the transport sector will be necessary, with an increase in EV usage being one of the primary methods to accomplish this (HKGOV LEGCO, 2021).

2. Infrastructures for Decarbonizing Electricity

This section mainly discusses key infrastructure technologies other than those already examined above. The decarbonization of electricity consumption involves three key strategies: (1) generating electricity within the region with non-fossil energy, such as nuclear and renewables; (2) importing non-fossil electricity from outside; and (3) decarbonizing fossil-fuel-fired electricity through CO_2 capture and storage. Hong Kong's local electricity generation infrastructure is almost entirely based on fossil fuels. The current plan to phase out coal-fired power plants by 2035 and replace them with natural gas-fired power plants can only reduce CO_2 emissions by about half, and the remaining half is a lot more challenging. Hong Kong's goal of carbon neutrality by 2050 largely depends on whether electricity generation can be entirely or at least deeply decarbonized. Three alternative pathways could exist, but all require substantial investments in new infrastructures.

2.1. *Generating carbon-neutral electricity locally*

Carbon-neutral electricity mainly comprises nuclear, hydropower, wind, and solar electricity, and these are crucial solutions for Guangdong, as discussed in Chapter 3 and the following sections. Politically and geographically, building a nuclear power plant in Hong Kong is nearly impossible due to its high population density and previous public resistance against nuclear energy. Hydropower requires a large river with enough elevation, which Hong Kong does not have, and Guangdong's potential is also very limited.

2.2. *Energy storage*

Energy storage is currently dominated by battery or pumped storage hydropower (PSH). PSH is a form of energy storage whereby water is pumped between upper and lower reservoirs (Tim, 2021). Traditionally, pumped hydro has been used as peaking power to provide power when demand is high. Water is released from an upper reservoir during high-demand times and pumped up during low-demand times (Tim, 2021). According to the International Hydropower Association (IHA), PSH must be doubled by 2050 to meet the Paris Climate Goals (IHA, 2020). PSH is the most significant form of energy storage, accounting for over 94% of global energy storage capacity (IHA, 2020). PSH offers greater flexibility to compensate for the intermittency and seasonality of other renewables (IHA, 2020). As of 2019, global PSH had a capacity of 158 GW, with an expected increase in capacity of 74 GW by the end of 2030 (IHA, 2020).

As discussed earlier, although daily energy demand is relatively static and can be plotted or predicted accurately, the energy supply from renewables cannot. Renewables such as solar and wind, by nature, are weather-dependent, with significant variations in daily output. An increase in storage is therefore needed to maintain electricity security and reliability. Currently, neither PSH nor battery power is likely to be able to store the necessary amounts of power required for Guangdong. One possible solution is through the utilization of smart grids and EVs. EVs connected to the grid can act as energy storage in emergencies or extreme-supply situations (Energy Storage and Transportation, 2021). Using intelligent controls to either slow down charging during peak demand or sell energy back to the grid can act as a revenue source, which is termed "vehicle to grid" (V2G).

Intelligent microgrids that analyze consumption patterns can optimize to achieve lower costs (Microgrid Knowledge, 2021). Small-scale experiments at Princeton do lend credibility to this idea. Princeton showed a 50% reduction in CO_2 emissions after using an intelligent microgrid compared to the typical grid (Siemens, 2020). Similar projects were done in California with 50 EVs using innovative charging control systems. In the California system, the cost of charging decreased by 15–30%, with the vehicles having a more standard charging time without peaks and troughs in energy supply compared to traditional charging (Grid Integration Group, 2020). EVs can primarily serve as storage systems since regular personal vehicles remain parked for 22 h daily (Shakeel and Malik, 2019). The use of V2G will likely be paired with the current fastest charging system, DC fast charging, which is rated at 90 kW and can charge a car in 20–30 min (Shakeel and Malik, 2019). As Guangdong and Shenzhen are big adopters of EV technology, this leads to ample growth opportunities in Guangdong.

2.3. *Importing carbon-neutral electricity*

Carbon-neutral electricity generated by China Southern Power Grid (CSG) mainly comprises nuclear, hydropower, wind, and solar electricity. The phase-out of natural gas-fired power plants in Hong Kong indicates that Hong Kong will rely on imported electricity, like all other land-scarce metropolitans in Mainland China and abroad. However, a large infrastructure gap exists in electricity transmission because Hong Kong's current electric grids have minimal interconnection with CSG (CLP, 2021). Furthermore, significant local energy storage capacities are necessary to optimize electricity supply and maintain current reliability.

More significant electricity usage within Hong Kong will lead to a correspondingly increased draw of power on the grid. Electrification of non-electrical systems within Guangdong will also lead to increased electricity consumption within the CSG network, which will likely require greater imports of electricity from neighboring provinces within the CSG and the China State Grid systems. Hong Kong's total end-use was 286,488 TJ, of which electricity comprised 161,709 TJ (56.44%) (Electrical and Mechanical Services Department, 2021). Of this, 44,571 TJ was imported into Hong Kong from Guangdong via the Daya Bay Nuclear Power Plant, with no electricity exports (HKGOV, 2020b). These numbers are slightly less than the generated amounts of electricity shown from other sources (177,033 TJ), as the 161,709 TJ figure excludes the electricity used by the power companies themselves as well as the energy lost from distribution (HKGOV, 2020b). This usage and loss amount is just under 9.5% of the total electricity produced. Correspondingly, Hong Kong had a surplus installation capacity of 12.25 GW in 2019 without including Daya Bay in Shenzhen, which exports most of its generation to Hong Kong (HKGOV, 2020b). The peak demand in 2019 was 9.6 GW. Over a 10-year time horizon, the maximum installed capacity in Hong Kong has decreased from 12.62 GW to 12.25 GW, while peak demand has fluctuated between 9.6 GW and 10.7 GW (HKGOV, 2020b). As a ratio of total power drawn in 2017, peak demand came closest to meeting the maximum installed capacity, with peak demand reaching 10.7 GW against a capacity of 12.3 GW, or 85.6% of the total installed capacity (HKGOV, 2020b). The complete electrification of Hong Kong would require a significant growth of Hong Kong's current electric generation capacity (excluding Daya Bay). Meeting these electrification numbers would require large-scale investments in new power plants and lead to increased CO_2 emissions from the electricity sector if the electricity were locally generated.

2.4. *CO_2 capture and storage*

CO_2 capture, utilization, and storage (CCUS) and CO_2 capture and storage (CCS) are processes in which CO_2 is captured from point-source emissions, such as coal and natural gas plants, cement production, or steel mills, and direct air capture (DAC). Then, CO_2 is compressed into a fluid state and injected over 800 m underground (Circular Carbon Economy, 2020b). Global underground storage is currently 260 million tons of CO_2, while conservative estimates are that over 4 trillion tons of CO_2 could be stored underground, which is over 121 times greater than the annual global emissions since 2019 (Circular Carbon Economy, 2020a). As of 2019, there were 51 large-scale (0.7–1 mta) CCS systems in various development stages, with 19 in operation and 4 in construction (CCS Institute Policy 2020). By August 2020, this had increased to 21 facilities in operation with 40 million tons of annual capacity and a further 3 in construction (Circular Carbon Economy, 2020b). By year-end 2020, 65 commercial CCS facilities were in various stages of development (Staff, 2020). If all currently planned CCS facilities were to be operated, total CO_2 capture would be 130 Mt/year, more than tripling current rates (IEA, 2020a). 2020 saw a 33% increase in CO_2 storage capacity compared to 2019, with significant growth in the United States, although CCS usage is still less than 1% of what the IPCC expects will be needed by 2050 to meet the Paris climate targets (Staff, 2020).

The primary advantage of CCUS is its ability to recover over 90% of CO_2 emitted from point-source emissions, such as coal plants (Handa and Baksi, 2020). India is also currently exploring the usage of CCUS from coal-powered industrial boilers (Handa and Baksi, 2020). Recent data from the Global CCUS Institute suggest that China's oil and gas fields alone would have storage potential for up to 8 GT of CO_2, while total storage potential could be over 3,000 Gt of CO_2 (GCI, 2020). The mitigation cost over a 30-year lifespan in Guangdong will be US$2 billion, or US$59/tCO_2, though changes in oil prices will affect this number (Wang *et al.*, 2020b). The IEA estimates that since the average age of coal plants in China in 2020 was only 13 years, up to 900 GW of coal could still be operational by 2050 (IEA, 2020c). However, retrofitting power plants' cost has decreased significantly from US$110 per ton of CO_2 captured in the 2014 Boundary Dam project to an expected US$45 for projects in 2025–2027. This may incentivize more plants in Guangdong to use CCS, especially in subcritical facilities (Stern and Berghout, 2021). In China, specifically for 2025, the costs regarding the post-combustion facilities (power plants) ranged between

230 and 310 yuan (US$35.5 and 47.7) in 2021 prices (Liu *et al.*, 2022).

In the iron and steel sector, using CO_2 capture systems raises costs by less than 10% compared to 35–70% when using electrolytic hydrogen (Stern and Berghout, 2021). This is a vital consideration, as Guangdong used 13.5 million tons of coal in smelting in 2019 and a further 5.1 million tons in coking (Guangdong Statistics, 2020). When using US-based case studies, the cost associated with CCS shows an increase in capital cost of 24% and O&M cost of 34% when building a new system, with a 30% total plant capture rate for a 650 MW plant (Sargent and Lundy, 2020). This would be associated with a 90% CO_2 efficiency rate and 33% of the flue gas associated with the plant (Sargent and Lundy, 2020). A full-scale CCS system with a 90% capture rate from the entire plant's expected capita cost was 60% higher, and O&M cost was calculated to be 47% higher (Sargent and Lundy, 2020). However, the actual CCS system in the 30% case was only 9.41% of the total capital cost and 21.1% in a fully integrated 90% capture system (Sargent and Lundy, 2020). CCS retrofitting could also help retain the over 22,000 jobs at Guangdong coal plants (Cui *et al.*, 2020). The retrofitting of plants may help create additional jobs since the continued operation of coal plants would allow coal mines to continue operating (Darrell, 2020).

In the United States, CCS is supported by tax incentives signed on December 27, 2020, for research and development of CCS and a two-year extension on tax credit applications until 2025 (Landry *et al.*, 2021). Industrial players have also begun investing in CCS, with Elon Musk promising US$100 million to the technology and ExxonMobil promising US$3 billion over five years (Landry *et al.*, 2021). If combined with bioenergy or used as a form of DAC, CCS is also considered a way to provide negative emissions (IEA, 2020b). Overall, CCS deployment is well below expectations, with the 2009 IEA report calling for 100 extensive CCUS facilities (300 Mt CO_2) by year-end 2020; however, the actual amount was only 40 Mt, or 13% (IEA, 2020c). Moreover, annual investment in CCS was only 0.5% of global investment in clean energy/efficiency technologies (IEA, 2020c).

Recent global developments during COVID-19 have pushed CCUS forward, especially in Norway and the United States (IEAGHG, 2022). Norway currently offers subsidy prices for CCUS of 80 euros per ton, with an expected rise of up to US$200 by 2025 (IEAGHG, 2022). This has helped increase the project development speed of both the Longship and Northern Lights CCUS projects, with the Northern Lights program having an expected capture capacity of up to 1.5 million tons per year by 2024 and a storage capacity of up to 5 million tons per year (IEAGHG, 2022). Further, the extension of the 45Q policy and subsidies provided by the Inflation Reduction Act of 2022 have increased the investment potential of projects in the United States (IEAGHG, 2022). CCUS projects can apply for subsidies starting at US$60 per ton of CO_2 for EOR projects and up to US$180 if using a combination of DAC and storage until 2033 (IEAGHG, 2022). This helped realize an additional 34 projects added to the pipeline within the United States (Global CCS Institute, 2022). Although it is too soon to say what the overall effect will be, these and other policy changes led to a 44% increase in CCS projects in development from 2021 to 2022, with a total of 196 projects in some form of development (Global CCS Institute, 2022).

CCUS with natural gas processing has 13 facilities currently operating, while the Boundary Dam Coal Plant in Canada has been running a 1 Mtpa CCS unit since 2014 (Global CCS

Institute., 2020). Studies conducted in 2020 suggest that 165 existing coal plants (175 GW) in China required CCS retrofitting to meet the 2°C goals and be stored in an average radius of 115 km (Wang *et al.*, 2020b). Of the existing coal plants, 513 GW have access to suitable storage, and 385 GW have storage locations within 250 km or less (Wang *et al.*, 2020b).To meet emission reduction goals 70% of natural gas and coal plants in China would require CCUS retrofitting to meet the 2°C climate goal (Wang *et al.*, 2020b). The IEA studies also show that northeast China is well suited for large amounts of CO_2 storage (Varro and Fengquan, 2020).

As of 2020, China had one large CCS facility in operation and two in construction, with three more in development (Global CCS Institute., 2020). This increased to two facilities in operation, two in construction, and four in development in 2022 (Global CCS Institute, 2022). Both operational facilities (CNPC Jilin Oil Field 0.6 Mtpa and Qilu Petrochemical 1.5 Mtpa) are used for enhanced oil recovery (EOR), with the remaining facility under construction for chemical production. Of the four large-scale sites in development, all are for power generation and in the early stages (Global CCS Institute, 2020b; Global CCS Institute, 2022). Under the Ministry of Science & Technology, China has developed a roadmap for deploying CCUS by 2030 (IEA, 2020c). The IEA's SDG predicts up to 400 Mt of CCUS in China by 2030 (IEA, 2020c). Over 2021–2022, one project was added to the pipeline, the Huaneng Longdon CCUS project, a 1.5 Mtpa project focused on the power generation sector (Global CCS Institute, 2022).

Guangdong has researched CCS's feasibility (Daiqing *et al.*, 2013; Ying *et al.*, 2016; Sun *et al.*, 2018; Zhou *et al.*, 2018c). Guangdong has only one CCS site, the Haifeng CCS test site located next to the Haifeng ultra-supercritical coal-fired power plant (2100 MW), with an annual capture rate of 20,000 tons, and hopes to scale it to a 1 million ton capture system (Global CCS Institute., 2020). In June 2022, Exxon, Shell, and CNOOC signed a MOU exploring the possibilities of building a CCUS hub project in Daya Bay, with CNOOC further exploring a storage site within the Pearl River Delta (Global CCS Institute, 2022). As fossil fuels will likely continue to be an essential part of the Guangdong energy mix with high numbers of large point source emitters, it is vital to understand how Guangdong can integrate CCS into current and future policies.

As there is no large-scale CCUS system in Guangdong, there is no CCUS infrastructure either. Infrastructure, especially in CCS transportation, is lacking worldwide. The United States, the largest CCUS user, had only 6,000 km of dedicated CCS pipeline as of 2012, compared to over 2 million km of natural gas pipeline (Global CCS Institute, 2012). However, other countries, such as Canada, have begun building new pipeline infrastructures far exceeding current capacity in preparation for expected projects coming online over the following years (Global CCS Institute, 2020c). For example, the new Alberta Trunk CCS pipeline has a capacity of 14.6 Mt of CO_2 per year, though current usage is only 1.6 Mt from two CCUS sources (Global CCS Institute, 2020c). Research has been conducted on how existing infrastructure, especially natural gas pipelines, can be converted to transport liquefied CO_2, especially to depleted natural gas or oil fields. China hopes to increase its natural gas pipeline length from 81,000 km in 2019 to 160,000 km by 2025 (Weiyuan, 2020). Costs could be lower to repurpose and reuse stranded assets in the oil and gas sector (World Energy Council, 2019). This is necessary for the Asia Pacific region, as the expected cost of decommissions within the oil and gas sector will be over US$100 billion (World Energy Council, 2019). The oil sector recorded over US$1 trillion in stranded assets (unusable assets, such as when oil fields dry up) between 1997 and 2017 (World Energy Council, 2019).

According to the Acorn project in Scotland, the repurposing of gas pipelines for CO_2 transport may cost as little as 1–10% of the cost of installing new pipelines directly (ACT, 2018), while by reusing the pipeline for CCUS, the value per kilometer is over five times higher than using the pipeline steel in other construction (ACT, 2018). Although maximum pressure varies, natural gas pipelines can handle the high pressures required in CO_2 transport if in liquid form or can be adapted to allow higher operating pressures. Of the three pipelines within the Acorn project, two already have a working pressure within 10 bar of what is needed for CO_2 transport (170 and 175 bar), requiring minimal retrofitting. Simultaneously, the third would also be suitable for CO_2 transport in a dense gas phase (ACT, 2018). Country-level studies back these findings, with, on average, the capture part of the CCS project requiring most (70–80%) of the investment cost (Ying *et al.*, 2016).

CO_2 from the CCS process must be transported and stored (Noh *et al.*, 2019). Besides pipelines, another way could be through the storage and transportation of CO_2 by ships (Noh *et al.*, 2019). A concept of a CO_2 terminal was proposed by Noh *et al.* (2019) for a 1 million ton per year site (2,740 tons per day) with a transport distance of 580 km. With two carriers, each carrier would be required to carry just over 8,800 tons per trip to be effective (three-day trip duration, with CO_2 kept at −27°C and 16 bar at all times), which can be accomplished with 9000 ton carriers (Noh *et al.*, 2019). Currently, the maximum size of storage tanks is 5,000 tons; as an idea, the terminal should have four tanks of 4,500 tons each or two tanks per carrier with two empty to allow for bad weather events as a buffer (Noh *et al.*, 2019). Compared to an LNG storage facility, the primary difference is the distance to CO_2 storage sites, which are much closer than LNG sites.

In contrast, LNG temperatures are much lower (−162°C) while pressure is 1 bar/atm (Noh *et al.*, 2019). This allows for much larger storage tanks at the terminal for LNG and slower loading times due to the size of the tanks and storage. A secondary case study considered the Nordic countries, where emissions from the steel and cement industry will need to be abated to meet emission goals (Kjärstad *et al.*, 2016). As distance increases, transport costs via pipeline increase linearly, while ship costs remain relatively stable as capital costs remain the same (Kjärstad *et al.*, 2016). The data suggest that ship-based transport is advantageous over pipelines for smaller volumes of 0.5–2 million tons (Kjärstad *et al.*, 2016). While for more significant amounts (5–20 Mtpa), pipelines hold an advantage up to distances of 730–1,180 km. Costs, when applied to real-world point-source emission locations, do show significant variations, from 4 euros per ton (5 Mtpa and 165 km) up to 25 euros per ton (1,025 km at 6.8 Mtpa) (Kjärstad *et al.*, 2016). Regulatory measures may make the Scandinavian CCS transport scheme more likely, as the EU approved transboundary shipments of CO_2 in 2020 (Global CCS Institute, 2020c). Transboundary shipments and multiple sources using the same pipeline infrastructure could significantly reduce the costs of future CCS proposals (IEA, 2020c). For example, estimates for a 180 km, 2.5 Mt/year pipeline show a transport cost of 5.4 euros per ton. However, the same pipeline length with a 20 Mt/year capacity (using a hub structure) would likely have a transport cost of only 1.5 euros per ton (IEA, 2020c). Using this shared infrastructure for transport would probably be helpful for China, as the country has multiple potential hubs (50 km from the storage location), with current emissions in the range of 50–300 Mt/year (IEA, 2020c).

Investment in decarbonization will lead to unavoidable stranded assets occurring along the entire supply chain. In coal, multiple power plants must first operate for reduced hours before shutting down entirely. Working with CCS in the gas and coal industries can prolong the expected lifetime of some asset classes as a transition technology. CCS can significantly decrease the carbon emissions associated with these technologies, allowing it to create a class of transition energy resources. These transition resources may be necessary to meet net-zero requirements, as the expected scale-up of other technologies, such as renewables, nuclear, and hydrogen, requires time to ramp up to meet the expected demand. For fuel types such as oil and gas, stranded asset costs will also include the investments made in the transport infrastructure to a much greater extent than coal. Coal as a commodity does not require special transport or refining by rail and can be stored outside in open sheds available for nearly immediate use. The primary stranded asset for coal is the power plant itself. For other fossil fuels such as oil and gas, stranded asset costs extend to the transport (pipelines), storage (LNG/oil terminals), and refinery capacity, which cannot easily be repurposed for other uses. Estimates on the total value of these sectors are hard to come by but will be a significant obstacle for CO_2 management in the long term as new builds will need to cease while current assets may be retired early.

3. Energy Infrastructure Gaps

After the initial announcement regarding China's 2060 net-zero goal, an analysis was conducted regarding the annual investment required to reach this target, revealing it to be about US$500 billion annually for 30 years (Monteith and Wang, 2020). This annual investment requirement would increase renewable generation between five- and eightfold, fully electrify passenger vehicles, and decrease associated emissions from industries and the built environment (Monteith and Wang, 2020). A generalized assumption based on population data would suggest Guangdong's fair share of the cost to be around 9% of the total, or US$45 billion per year. However, as mentioned previously, Guangdong is not energy self-sufficient, importing significant power from neighboring less-populated provinces through the CSG grid system (CSG, 2019). Further, as previously established, to meet global net-zero goals, CCUS will be required both in the form of DAC and point-source capture (Global CCS Institute, 2022). Previous estimates on the cumulative amount of removal required to stay within IPCC goals have ranged from 333 billion tons to over 1.2 trillion tons, with the GCCSI also holding a range between 511 billion and nearly 1.3 trillion tons (Global CCS Institute, 2022).

Assuming that the capacity factors in Table 2 in Chapter 2 remain constant, available and planned power plants in Hong Kong, Macau, and Guangdong will generate 747 TWh of electricity in 2025. The electrification rate of Guangdong's end-use energy consumption was 35.9% in 2020, while in Hong Kong, the rate was 58.4% (Guangdong Bureau of Statistics and National Bureau of Statistics, 2022). Electrification and further economic growth will increase electricity consumption, while energy conservation and efficiency will reduce it. As a conservative estimate of the energy infrastructure gap for the region to achieve carbon neutrality, we assume that electricity generation within the region will double in 2060 from the 2025 level (Figure 7). The calculated electricity generation took only 10 years to double from 2015 to 2025. The annual growth rate is assumed to be faster in the first decade from 2016 to 2035 and slow down until 2060.

Under the current capacity amount, Guangdong and Hong Kong have 83 GW of coal under development or in operation, 61.6 GW of natural gas, 39 GW of nuclear, 31 GW of wind, and 28.2 GW of solar, with much of the stated capacity for wind and solar projects still under development with uncertainties over project completion (GEM, 2023). Total fossil-fuel-based capacity (planned and operational) stands at 144.6 GW vs. 98.2 GW of non-fossil-fuel-based capacity, although much of this is still in the planning stages (GEM, 2023). Further, Guangdong in

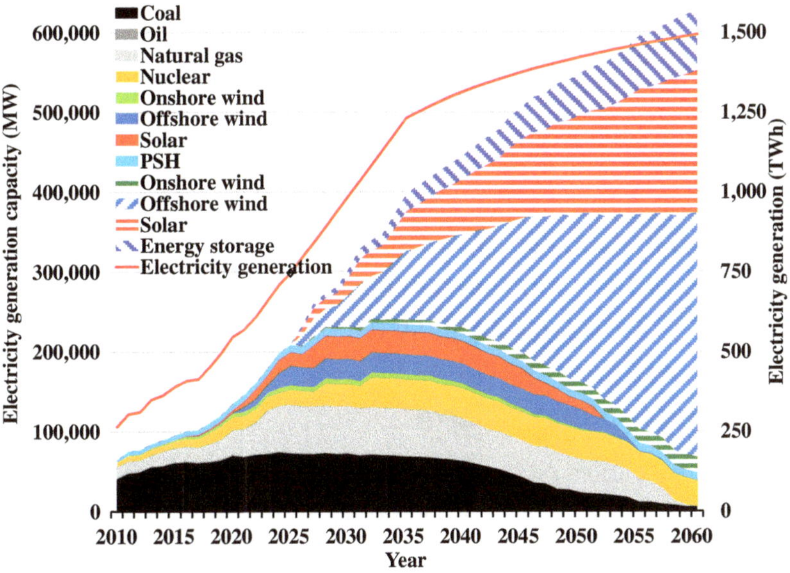

Figure 7. Electricity generation capacities in Hong Kong, Macau, and Guangdong Province in the Renewables Scenario (unit: TWh) (solid colors refer to those in operation or planned to be in operation and to be retired after 2022; dashed colors indicate those necessary additions for meeting expected electricity demand).

2019 only generated a little over half of its electricity (343.4 TWh compared to 669.6 TWh), with Yunnan providing 200 TWh via hydropower (CSG, 2020). Utilization rates, as previously mentioned, for wind and solar are significantly lower (1,800 and 2,082 h annually, respectively) than for thermal and nuclear power plants, at under half for thermal plants (4,293 h) and under a third (7,394 h) when compared to nuclear plants (CGN, 2020).

In estimating future energy infrastructure gaps, three distinct scenarios are worked out for Guangdong Province and Hong Kong to approach carbon neutrality. Power plants that have been planned will be constructed as scheduled. They will also be retired to reduce electricity generation when they reach the expected lifetime. To meet the gap in local electricity generation in future years, the first scenario exclusively relies on wind turbines and solar PV (thus referred to as the "Renewables Scenario" hereafter). The second scenario adopts only coal- and natural gas-fired power plants with CCS (thus referred to as the "CCS Scenario" hereafter). The third scenario utilizes only nuclear power plants for local electricity generation (thus referred to as the "Nuclear Scenario" hereafter). More scenarios could be constructed by combining the three pathways.

3.1. *Renewables Scenario*

In the Renewables Scenario, the gap in local electricity generation will be filled only by wind and solar capacities from 2026 to 2060. During 2021–2025, our data on existing and in-pipeline projects show that the region adds 1,476 MW of onshore wind, 23,164 MW of offshore wind, and 13,497 MW of solar PV (Figure 7). We assume that their relative ratios (3.9%, 60.7%, and 35.4%, respectively) will remain constant with future annual capacity growth. Due to offshore wind's significantly higher capacity factor (Table 2 in Chapter 2), their relative ratios of electricity generation will be 3.6%, 77.5%, and 18.9%, respectively. Furthermore, battery-based energy storage capacity in MW is assumed to be 15% of these new wind and solar power capacities, and the electricity storage duration is expected to be 3 h.

Accordingly, in 2060, these additional capacities of onshore wind, offshore wind, solar, and battery-based energy storage in the region are expected to be 19.5, 306.2, 178.4, and 75.6 GW, respectively (Figure 7). Due to the retirement of most fossil-fuel-fired power plants, less than 1% of local electricity generation capacity would be fossil-fuel-fired.

From the perspective of local electricity generation, nuclear electricity will play a vital role (17.1% of the total), even though the

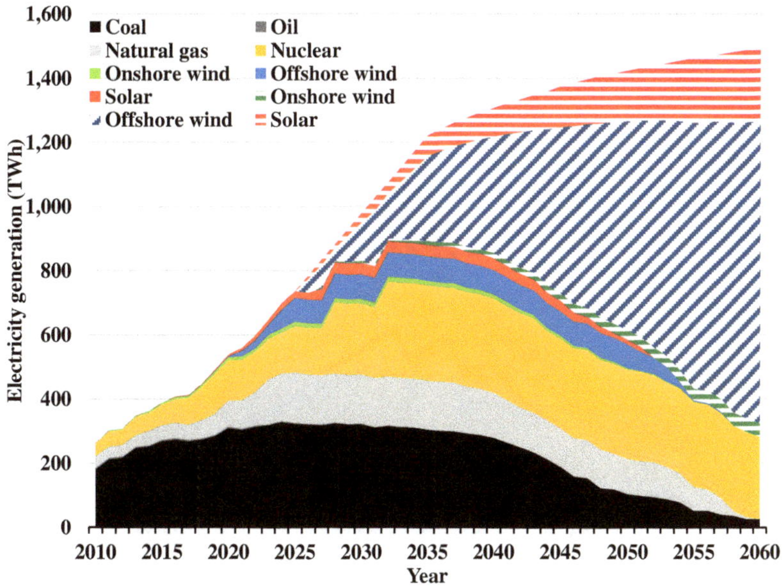

Figure 8. Electricity generation by fuel in Hong Kong, Macau, and Guangdong Province in the Renewables Scenario (unit: TWh).

Renewables Scenario does not add new nuclear power plants other than those already planned. The shares of onshore wind, offshore wind, and solar electricity will be 2.9%, 62.9%, and 15.3%, respectively (Figure 8). Fossil-fuel-fired electricity will be responsible for less than 2% of the total, indicating the approach toward carbon neutrality (Figure 8).

The asset value of these energy infrastructures will be US$169.9 billion in 2025 and will reach US$324.4 billion in 2060 (Figure 9). The structure will also be transformed. In 2025, the shares of fossil-fuel, nuclear, renewables, and energy storage infrastructures will be 59.3%, 17.3%, 20.3%, and 3.1%, respectively. In 2060, their shares will become 2.0% (including 0.2% for coal-fired power plants, 0.5% for LNG terminals, and 1.4% for oil refineries; the numbers do not precisely add up due to rounding errors), 7.4%, 77.6% (including 2.6% for onshore

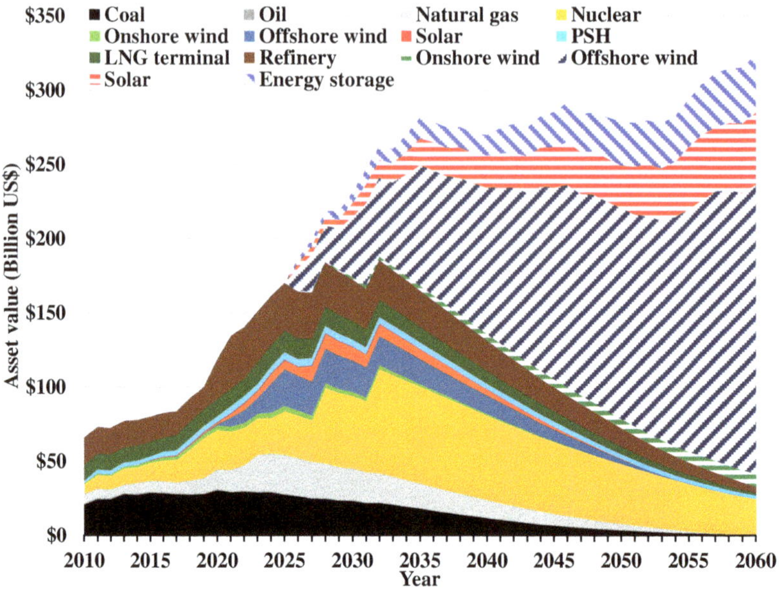

Figure 9. Asset values of energy infrastructures in Hong Kong, Macau, and Guangdong Province in the Renewables Scenario (unit: billion US dollars).

wind, 60.0% for offshore wind, and 15.0% for solar PV), and 12.9% (including 0.8% PSH and 12.1% battery-based energy storage), respectively (Figure 9).

Assuming that the geographical distributions of additional onshore wind, offshore wind, and solar PV in the Renewables Scenario follow the identical distributions of those coming online during 2021–2025 (Figure 4 in Chapter 3), we can estimate the geographical distribution of assets in the coming decades. In 2060, those existing and planned projects as of 2023 only account for 10.2% of the assets because most would have been depreciated or retired (Figure 10). Non-GBA Guangdong will have US$270.3 billion, or 83.3% of all assets, in 2060, US$247.0 billion of which are newly added renewables. GBA Guangdong will have US$52.7 billion, or 16.3%, in 2060. In sharp contrast,

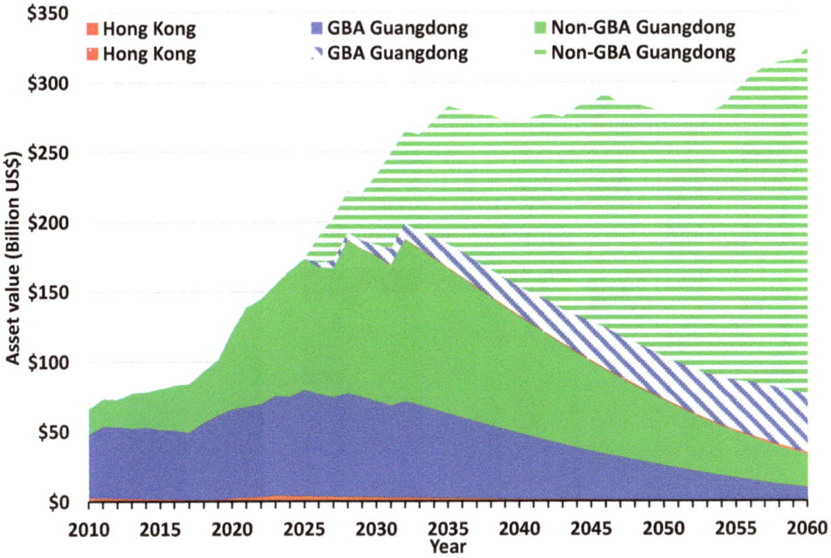

Figure 10. The geographical distribution of energy infrastructure asset values in the Renewables Scenario (unit: billion US dollars).

non-GBA Guangdong was responsible for US$55.9 billion, or 45.8% of all energy infrastructure assets, in 2020, while GBA Guangdong had US$63.4 billion, or 51.9%, in 2020 (Figure 10). As a result, the Renewables Scenario will witness a dramatic shift in the energy economic landscape, with energy infrastructure assets being distributed from the more developed GBA Guangdong to the less developed non-GBA Guangdong. This outcome could alleviate the substantial regional disparities that have long persisted in Guangdong Province.

Due to the retirement and depreciation of energy infrastructures, although this analysis does not consider inflation in estimating future investment, the total required investment will be significantly higher than the asset values in 2060. From 2026 to 2060, an additional sum of US$611.5 billion is expected to be invested, including US$16.6 billion of onshore

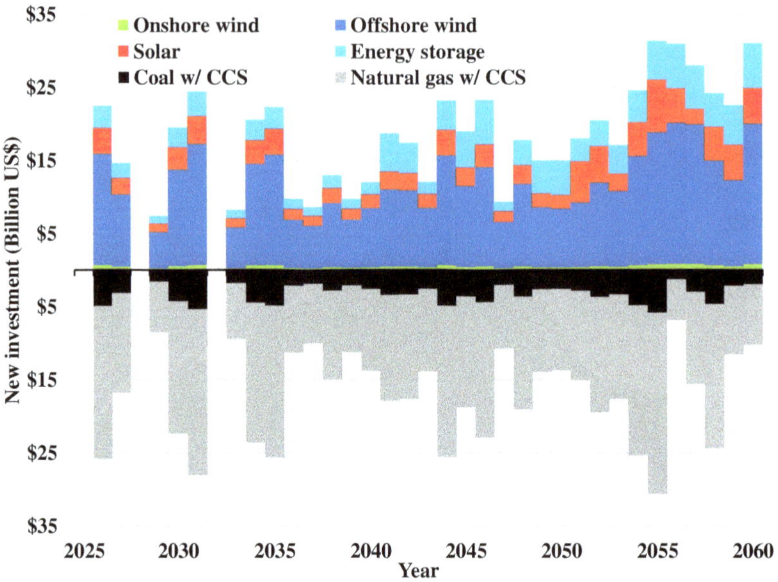

Figure 11. Annual new investment in the Renewables and CCS Scenarios (unit: billion US dollars).

wind, US$380.1 billion of offshore wind, US$100.5 billion of solar PV, and US$114.3 billion of battery-based energy storage (Figure 11). The annual required investment has a slightly increasing trend over the 35 years, with the average being US$17.5 billion per year.

3.2. *CCS Scenario*

In the CCS Scenario, the gap in electricity generation will be filled only by coal-fired and natural-gas-fired capacities from 2026 to 2060, all equipped with CO_2 capture and storage. During 2021–2025, our data on existing and in-pipeline projects show that the region adds 5,800 MW of coal-fired power plants and 28,915 MW of natural gas-fired ones (Figure 12). Their ratio, 16.7% vs. 83.3%, is assumed to be constant in the future addition

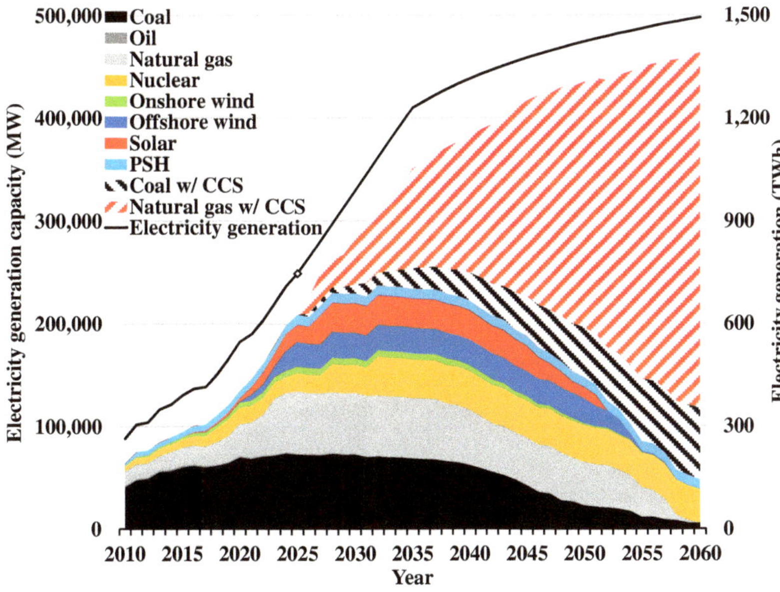

Figure 12. Electricity generation capacities in Hong Kong, Macau, and Guangdong Province in the CCS Scenario (unit: MW).

of capacities. Because coal-fired power plants have a higher capacity factor on average, they are responsible for 25.4% of additional fossil-fuel-fired electricity generation.

In 2060, 69.3 GW of coal-fired power plants and 345.4 GW of natural gas-fired power plants will be required in the region. Due to their high dispatchability, no additional energy storage infrastructures are assumed to match electricity generation and demand. The CCS Scenario is designed to be essentially the opposite extreme of the Renewables Scenario. Fossil-fuel-fired power plants will be responsible for 90.8% of total electricity generation capacity, 82.9% of total electricity generation, and 92.1% of total energy infrastructure asset values in 2060 (Figures 12–14). Because CCS can avoid about 85% of CO_2 emissions, the CCS Scenario will accordingly witness much greater residual CO_2 emissions than the Renewables Scenario.

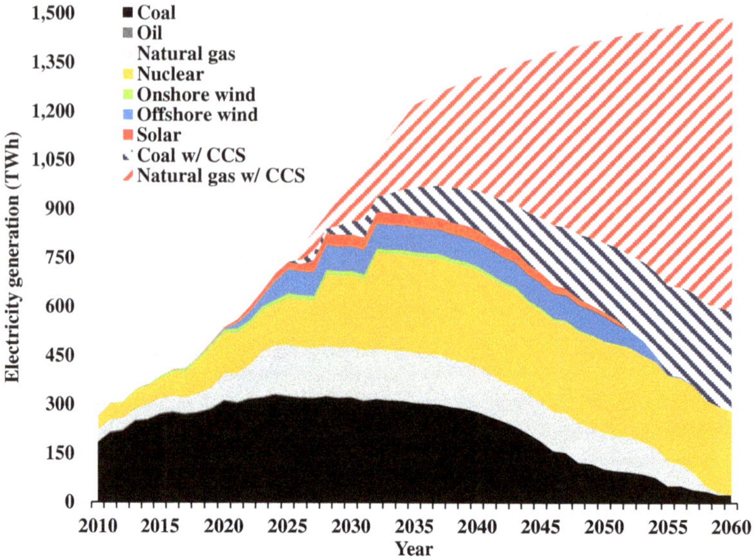

Figure 13. Electricity generation by fuel in Hong Kong, Macau, and Guangdong Province in the CCS Scenario (unit: TWh).

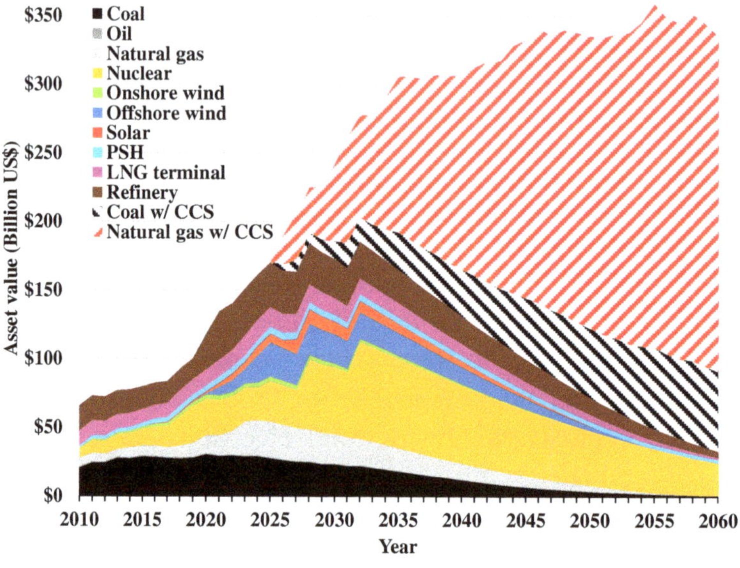

Figure 14. Asset values of energy infrastructures in Hong Kong, Macau, and Guangdong Province in the CCS Scenario (unit: billion US dollars).

Similar to the case of the Renewables Scenario, we assume that the geographical distributions of additional fossil-fuel-fired power plants in the CCS Scenario also follow the identical distributions of those coming online during 2021–2025 (Figure 4 in Chapter 3). As a result, non-GBA Guangdong will hold US$99.1 billion in energy infrastructure asset value (29.5% of the total), including US$75.7 billion of newly added fossil fuel infrastructures during 2026–2060. GBA Guangdong's respective numbers are US$215.9 billion and US$206.2 billion. This marks another sharp contrast between the two scenarios, as the CCS Scenario further strengthens the economic importance of GBA Guangdong.

The CCS Scenario corresponds to the value of a similar asset in 2060, being US$335.5 billion, while total investment during 2026–2060 will be US$572.4 billion, including US$109.4 billion

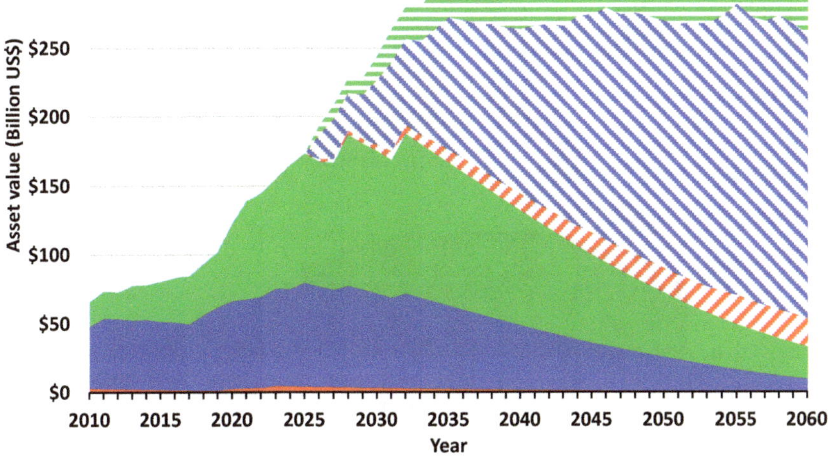

Figure 15. The geographical distribution of energy infrastructure asset values in the CCS Scenario (unit: billion US dollars).

for coal-fired power plants with CCS and US$462.9 billion for natural gas-fired power plants with CCS (Figure 11). On average, US$16.4 billion per year should be invested (Figure 15).

3.3. *Nuclear Scenario*

The Nuclear Scenario only relies on nuclear power plants to fill the gap in local electricity generation from 2026 to 2060. Given its very high-capacity factor and low despicability, we also assume that battery-based energy storage infrastructures will be added to reach 30% of the newly added nuclear capacity, and the storage duration will be 4 h.

In 2060, 155.2 GW of nuclear power plants and 46.6 GW of battery-based energy storage capacity will be needed (Figure 16).

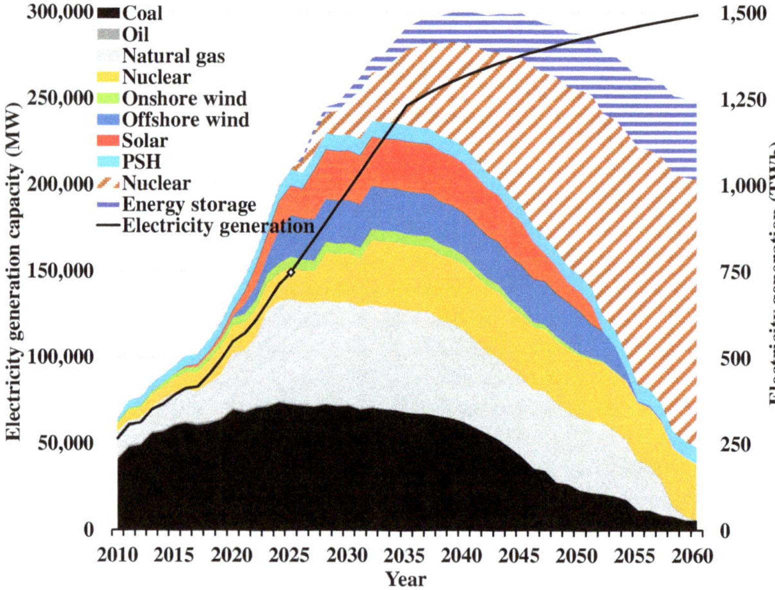

Figure 16. Electricity generation capacities in Hong Kong, Macau, and Guangdong Province in the Nuclear Scenario (unit: MW).

The Nuclear Scenario requires 250.2 GW of total electricity generation capacity in 2060, much lower than 628.2 GW in the Renewables Scenario and 463.0 GW in the CCS Scenario. The primary reason for the differences is nuclear power plants' much higher capacity factors (Table 2 in Chapter 2).

In the Nuclear Scenario, in 2060, nuclear power plants will account for 75.2% of total electricity generation capacity, while energy storage will occupy 22.5%, including 3.9% of PSH and 18.6% of batteries (Figure 16). Fossil fuels will form the remaining 2.4%. Because energy storage capacities do not generate electricity, electricity generation becomes almost entirely nuclear, with its share being 98.2% (Figure 17). Due to the meager share, 1.8%, of fossil-fuel-fired electricity, local electricity generation in Hong Kong, Macau, and Guangdong Province will be nearly

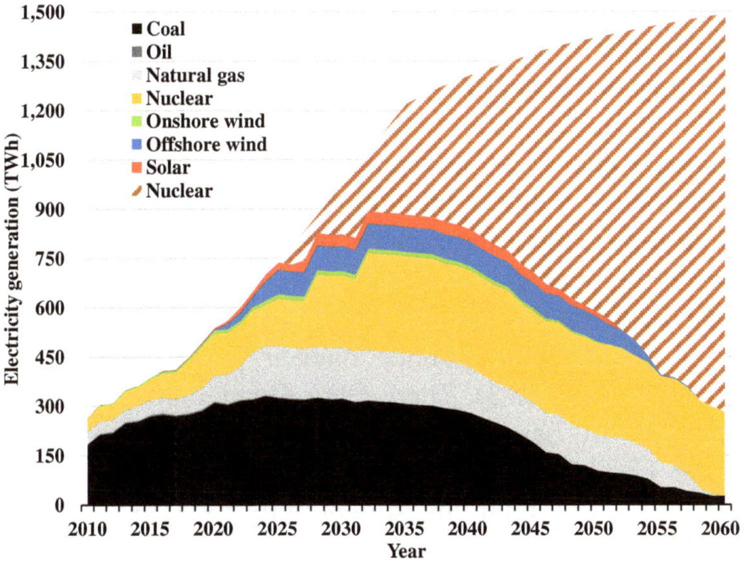

Figure 17. Electricity generation by fuel in Hong Kong, Macau, and Guangdong Province in the Nuclear Scenario (unit: TWh).

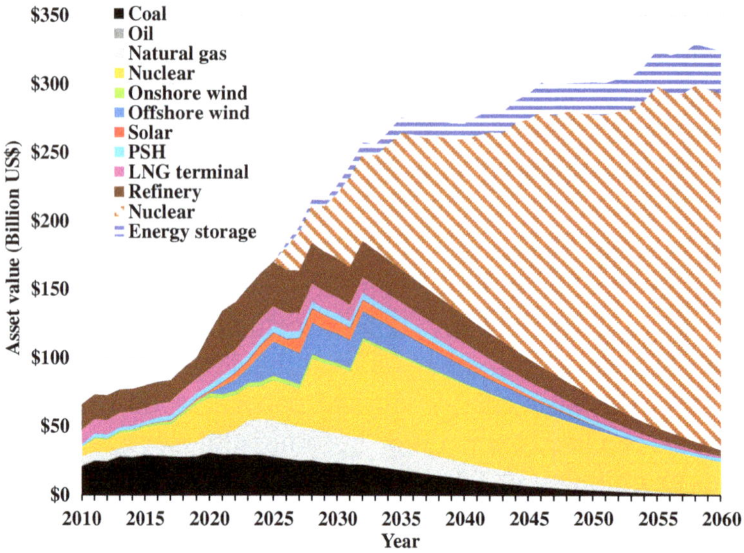

Figure 18. Asset values of energy infrastructures in Hong Kong, Macau, and Guangdong Province in the Nuclear Scenario (unit: billion US dollars).

carbon-neutral in the Nuclear Scenario, similar to the situation in the Renewables Scenario.

The asset value of those accounted for energy infrastructures will be US$324.6 billion in 2060, including US$291.5 billion of newly added energy infrastructures during 2026–2060 (Figure 18). The geographical distributions of new infrastructures in the Nuclear Scenario are assumed to follow the identical distributions of existing and planned nuclear power plants in 2030 (Figure 4 in Chapter 3). Then, GBA Guangdong and non-GBA Guangdong will have 48.7% and 51.3% of total asset values in 2060, or US$158.1 billion and US$166.4 billion, respectively (Figure 19). The total investment will be US$480.2 billion, US$388.1 billion of which is for nuclear power plants (Figure 20). On average, US$11.1 billion will be invested annually in constructing nuclear power plants and another US$2.6 billion in battery-based energy storage (Figure 20).

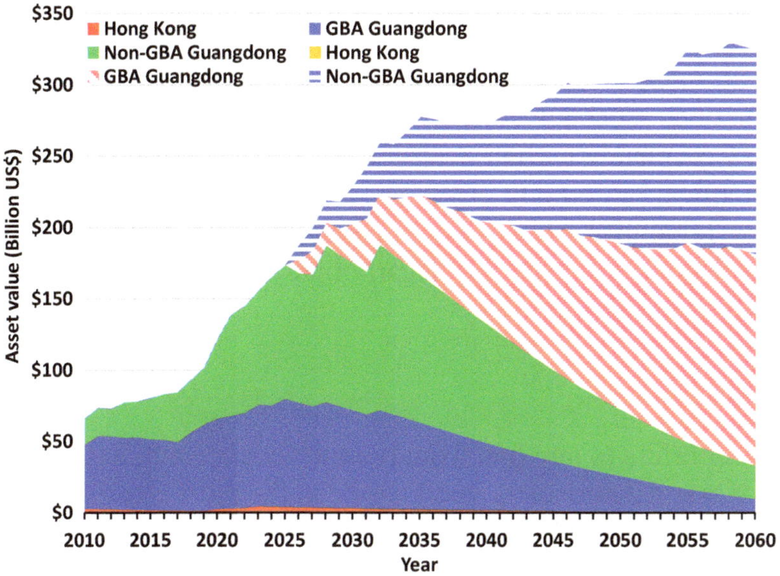

Figure 19. The geographical distribution of energy infrastructure asset values in the Nuclear Scenario (unit: billion US dollars).

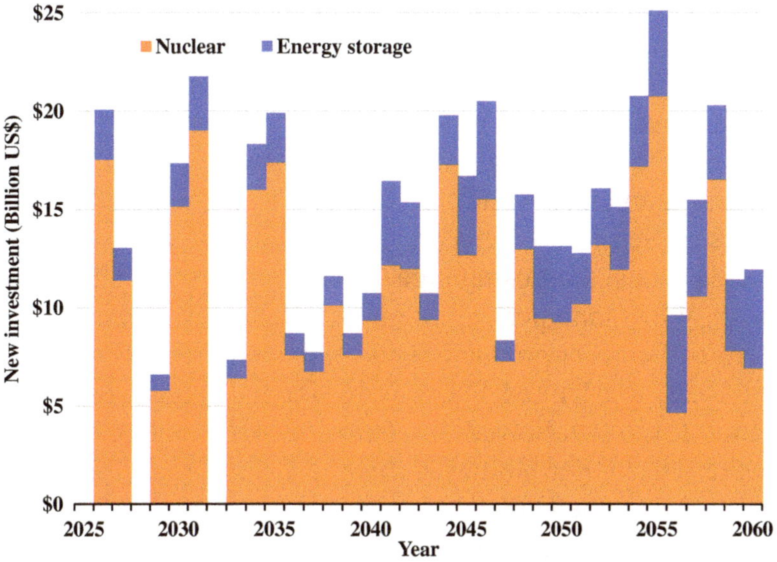

Figure 20. Annual new investment in the Nuclear Scenario (unit: billion US dollars).

4. Neutralizing Remaining CO_2 Emissions

Not all energy services can be technologically or economically electrified, especially those of ocean-going shipping and aviation. Completely carbon-neutral electricity may not be the most economical option due to various resource constraints and electricity reliability and resilience concerns. This section explores several other means to decarbonize these remaining CO_2 emissions.

4.1. *Biofuel*

Biofuel refers to any fuel from biomass, i.e., plant, algae, or animal waste. Historically, wood has been the most commonly used type of biofuel and is still used to produce heat and for cooking in rural low-development communities (Selin and Lehman, 2022). Biofuel, in a more recent context, usually refers to ethanol-based fuels, often created through the fermentation of sugar and sugar cane (Selin and Lehman, 2022). Biodiesel is produced primarily from soybeans and oil palms (Selin and Lehman, 2022). In the United States, corn is also used to create a gasoline/ethanol blend (90/10) called gasohol (Selin and Lehman, 2022). In recent years, biofuel production has moved away from using food crops toward using products such as wood chips, crop residue, and municipal waste for ethanol and waste cooking oil for biodiesel (Selin and Lehman, 2022). Current research in biodiesel is focused on using algae, as it could potentially use significantly less land per unit of output (Selin and Lehman, 2022). In terms of current production and expected production amounts until 2028, the United States is by far the largest producer of ethanol (50%), followed by Brazil (24%), and China (8%) (Seabra, 2021). For biodiesel, there is a slight variation between the EU (36%), the United States (19%), and Brazil (12%). China is expected to be the eighth-largest producer, with a market share of 3% (Seabra, 2021). According to the IEA, growth in biofuel is expected to be 28% (41 billion liters) between 2021 and 2026

(IEA, 2021), while an accelerated growth case would see production of liquid biofuels peaking at 64% above the 2020 levels (9% yearly growth) (IEA, 2021). However, even this would be below what is predicted to be required under global net-zero goals, requiring more than a doubling of biofuel production, with an annual compound growth rate of 15% (IEA, 2021).

Biofuel could be the most convenient way to decarbonize unelectrified energy consumption. Bioethanol and biodiesel could fit into the current infrastructure with relatively insignificant changes compared to other alternative methods. Research shows that on a global level, land usage increased by 25 million hectares, of which 13.5 million hectares were for net biofuel production (no co-benefits) (Seabra, 2021). Biofuels also provided around 3.44% of global transport fuel demand in 2018 (Seabra, 2021).

Growth is also expected to increase significantly in ethanol production. However, the global production of biofuel so far is limited, and its high demand for land and freshwater resources also heavily constrains its role in the carbon neutrality future of Hong Kong and the GBA. For example, Hong Kong uses 49 km^2, or 4.4%, as agricultural land and a further 12 km^2, or 1.4%, for fish ponds (Land Department, 2022). Increasing the amount of land used for agriculture and, in turn, biofuel would require decreasing the current amount of land acting as carbon sinks, i.e., woodland (26%), shrubland (22.8%), and grassland (16.1%) (Land Department, 2022). For Guangdong, agriculture also does not seem to be a significant priority of the government. Although the value added increased from 100 billion RMB to 448 billion RMB from 2000 to 2019, the actual area under cultivation decreased by 15.5% in the same period (Guangdong Bureau of Statistics and National Bureau of Statistics, 2020). The economic value primarily seems to be due to the effect of inflation and the increasing value per unit of goods. This could be attributed to the fact that the production of grain-based crops and sugarcane peaked in the 1980s and 1990s (Guangdong Bureau of Statistics and National Bureau of Statistics, 2020). For ethanol production, Guangdong will likely need to increase import of the fuel if it plans to use gasohol, as current sugarcane production is only 60% of its peak in 1992 (Guangdong Bureau of Statistics and National Bureau of Statistics, 2020). However, oil-based crop growth has increased slightly (Guangdong Bureau of Statistics and National Bureau of Statistics, 2020). It is, therefore, unlikely that Guangdong or, by extension, the GBA can significantly improve their biofuel production due to the decrease in agriculture area and the recent decrease in forested area, from which an increased agriculture area would likely come (Guangdong Bureau of Statistics and National Bureau of Statistics, 2020).

Biofuel could play a complementary role in decarbonizing shipping and aviation as a sustainable marine fuel (SMF) or a sustainable aviation fuel (SAF), respectively. Under current International Maritime Organization (IMO) rules and regulations, the shipping industry seeks to create serious efficiency increases, with a goal of 50% reduction of CO_2 per ton-mile compared to the baseline of 2008. Individual companies have made more significant pledges to decarbonize faster. Maersk, the world's second-largest shipping company, recently announced its net zero by 2040 goals while pledging to be on track to meet IPCC 1.5° goals by 2030. Another way shipping companies seek to decrease emissions, aside from hydrogen-based fuels, is through biofuels. Biofuels have already been used in land-based vehicles, such as cars, through synthesis and ethanol-based fuel. However, biofuel usage in the shipping industry is still in its infancy. In January 2021, of the over 2 billion tons of restaged shipping, only 18 ships with a combined tonnage of just over 330,000 tons ran on pure biofuels, with an additional four vessels utilizing a mixture of blended fuel types.

4.2. *Hydrogen*

Hydrogen, primarily from natural gas reforming, has been used for years for agriculture and heating. The hydrogen in ammonia is a primary ingredient for many nitrogen-based fertilizers ('Global Hydrogen Review 2021', 2021). Blue hydrogen, or hydrogen produced from fossil fuels and CCS, would constitute a move away from current production, which uses natural gas to produce hydrogen (gray hydrogen) and has a higher emission intensity of about 10–12 kg of CO_2 per kg of hydrogen produced (Taylor, 2020). The transition from gray hydrogen to blue hydrogen and yellow hydrogen (nuclear) is seen as an essential step in the energy transition, with blue hydrogen as a step toward creating the infrastructure required for green (renewable) hydrogen production (White, 2019). Hydrogen is expected to grow as an industry, primarily as a clean fuel source used in industry or heating. Demand in a carbon-friendly scenario will increase to 30 Mt of blue hydrogen and 25 Mt of green hydrogen by 2030 (IRENA, 2020). Hydrogen can also be incorporated into current infrastructures. For example, many European homes can handle a 20% hydrogen and 80% natural gas mix without difficulty in current appliances, decreasing carbon intensity (Taylor, 2020).

In Hong Kong, hydrogen is heavily used by Towngas, with 46.3–51.8% of town gas being hydrogen (Towngas, 2019). However, the production of town gas in Hong Kong is dominated by a natural gas (61%) and naphtha (38%) fuel mix (Towngas, 2019). Hydrogen storage can have a relatively cheap cost of around US$0.23/kg, at least in a gaseous state within salt caverns, with the price expected to halve over time (SNAM, 2020). At the same time, CLP and CSG have signed agreements to consider the possibility of converting the Black Point Power Plant to operate on 100% hydrogen (HKGOV, 2021a).

The Chinese government hopes hydrogen could account for 10% of its energy system by 2040, with demand expected to reach 60 million tons by 2050 (Brasington, 2020). The most significant increase in hydrogen usage is likely to be in steel manufacturing since, even without making substantial changes to iron furnaces, up to 35% of the natural gas used can be substituted with hydrogen, with demonstration trials projected for 2025 (Bermudez *et al.*, 2020). Hydrogen will also likely be used in the transport sector, primarily through the car market, as both EVs and hydrogen cars will likely increase in the market (Brasington, 2020). Hydrogen vehicles, for example, would be used as hydrogen fuel cells powering an EV. However, the overall efficiency would be twice that of a standard diesel or gasoline vehicle (Wind, 2016). High energy loss (50% of total produced) from production to delivery at the fuel station limits hydrogen effectiveness (Wind, 2016). At the same time, the pressure in a H_2 tank, due to the small size of the vehicle, needs to be high, up to 700 bar (300 bar for buses), to provide enough range (Wind, 2016). Multiple projects are underway to reduce costs, with more advanced hydrogen models (Toyota Mirai 2015) storing over 5 kg of hydrogen and offering up to 700 km of range and a max speed of 178 km/h (Wind, 2016).

Green hydrogen could be produced in countries and regions with great renewable energy potential and then transported to consumption centers where local production of green hydrogen is constrained. Instead of building long-distance, high-voltage electricity transmission lines, hydrogen could be transported as stored energy. If hydrogen were made a pathway for achieving carbon neutrality in Hong Kong and the GBA, the region would require infrastructure to import, transport, and store hydrogen, besides other facilities to consume hydrogen.

Fundamentally, the challenges involved in storing and transporting pure hydrogen involve the need for either large amounts of space, extremely low temperatures, very high pressures, or preferably all three (Andersson and Grönkvist, 2019). At standard room conditions of 20°C and 100 bar, the density of hydrogen is only 7.8 kg/m^3 (Andersson and Grönkvist, 2019). Storage containers can reach pressures of up to 700 bar, but building such containers increases the cost significantly due to the material used in construction (Andersson and Grönkvist, 2019). At the same time, liquid hydrogen must be collated to −253°C to be kept in tanks of pressures in the range of 6–350 bar (Nash *et al.*, 2012). Furthermore, although it weighs about three times more, hydrogen is far less energy-rich by volume, especially in its gaseous state (Nash *et al.*, 2012). Hydrogen liquefaction plants are also generally relatively small, with the largest capacity of only 34 tons per day, although most plants today are within the 1–10 tons range (Andersson and Grönkvist, 2019). The energy required per kg of hydrogen produced is about 10 kWh, although experiments suggest that larger plants could bring this down to below 6 kWh per kg (Andersson and Grönkvist, 2019). However, the capital cost of liquefication plants is still high, with even a theoretical 100-ton-per-day plant incurring 40–50% of the project cost in construction (Andersson and Grönkvist, 2019).

Hydrogen can either be transported as a liquid or in its gaseous state (Gerboni, 2016). Currently, tubes/cylinders (usually made of a steel alloy) can be used as containers to hold hydrogen, usually at pressures of 180–250 bar (Gerboni, 2016). As a liquid, up to 4 tons can be transported per container either by rail or truck (Gerboni, 2016). Pipelines may also be used for storage or transport, with pressure and diameter similar to a natural gas line, storing 12 tons of hydrogen per km (Andersson and Grönkvist, 2019). Most hydrogen pipelines' operating pressure is 10–20 bar, and the diameter is about 25–30 cm (Gerboni, 2016). Hydrogen pipelines are about 10% more expensive than conventional natural gas lines (Gondal, 2016). Compression stations are also required and usually placed between 80 and 100 km apart, with each station using 3–5% of the gas transported as fuel (Gondal, 2016). However, smaller pipe diameters could see substantial decreases in costs as the ability to use different materials becomes available (Gerboni, 2016).

Hydrogen may also be transported by ships, with extensive research conducted in the 1980s and 1990s (Gerboni, 2016). A Japanese study (1993–2020) looked at other proposals for ships capable of transporting up to 200,000 m^3 of liquid hydrogen (Gerboni, 2016). More recent results show that ships carrying 160,000 m^3 may be possible, using 0.2% of the total fuel daily to power themselves (Gerboni, 2016).

The move toward hydrogen ships is significant since Hong Kong sells vast amounts of ship fuel (Jiang *et al.*, 2020). Ammonia may also be used as a ship fuel (Reuters, 2021). According to Citibank, the market for renewable ammonia could be a 6 billion euro industry by 2030 (Reuters, 2021). CF Industries believes that its new plant in Louisiana could produce 18,000 tons of green ammonia by 2023, 450,000 tons in 2026, and 900,000 tons by 2028 (Reuters, 2021). The advantages of ammonia over liquid hydrogen are its higher energy density, which nearly double that of liquid hydrogen by volume, and that it is easier to ship and distribute (Service, 2018).

Hydrogen use in shipping has increased in importance. Much interest in hydrogen and hydrogen-based products comes from the IMO and internal net-zero goals of large shipping companies, such as Maersk, MSC, and COSCO (UNCTAD, 2021). Hydrogen utilization has seen significant investment, with US$37 billion in pledged public money and US$300 billion in promised private financing. However, to meet net-zero goals for the hydrogen sector, the investment must reach up to US$1.2 trillion between 2030 and 2050 (Hydrogen Council, 2020). For the shipping industry to meet the IMO goals of a 50% decrease in emissions per ton-km from the 2008 baseline, it will require a yearly investment of US$40–60 billion from 2030 to 2050 (US$800–1.2 trillion) (Hydrogen Council, 2020).

Ammonia would fuel up to 45% of commercial shipping to achieve the net-zero goal, with total hydrogen demand in the transport industry reaching 100 million tons. This amount is currently much greater than the total hydrogen produced globally and more than the 20,000 tons used within the transport industry ('Global Hydrogen Review 2021', 2021). However, current goals and regulations to meet these amounts would require a considerable ramp-up, with currently pledged projects showing that 1% of fuel consumption in the shipping industry will come from hydrogen in 2030 and 8% from ammonia-based projects ('Global Hydrogen Review 2021', 2021). As such, current at-scale hydrogen-based shipping projects are limited. Norway started the first hydrogen-powered shipping connection in 2020, while China began manufacturing an ammonia-based ship in 2021 for Greece's Avni International. The final byproduct of hydrogen-based fuel currently in development is methanol (CH_3OH), which, although hydrogen-based, still has carbon within it but is less polluting than traditional hydrocarbons, such as heavy bunker fuel oil ('Global Hydrogen Review 2021', 2021).

4.3. *Nature-based carbon sinks*

Remaining CO_2 emissions can be offset by nature-based carbon sinks. Globally, the passing of the Kunming–Montreal Global Biodiversity Framework (GBF) in 2022 has increased the profile of nature-based carbon sinks and biodiversity in general (UNEP, 2022). The GBF seeks to increase the proportions of land and oceans protected from 17% and 8%, respectively, to 30% each by 2030, with the restoration of an additional 30% of ecosystems that have been degraded (UNEP, 2022). As such, further mobilization of funding for the conservation efforts is expected, including the phase-out and reforming of subsidies harming biodiversity (US$500 billion annually), increasing public and private funding of biodiversity projects to US$200 billion per year, and increasing financing transfer from developed to developing countries regarding biodiversity to US$30 billion per year (UNEP, 2022). Under China's 14th Five-Year Plan (2021–2025), the central government expects to increase forest coverage from 23% in 2020 to 24.1% in 2025, equating to 36,000 km^2 annually (Climate Action Tracker, 2023).

Carbon sinks in the context of Guangdong refer to the amount of carbon the province's vegetation can sequester (take in) over a given year. This is usually seen as forests, of which Guangdong has plenty, with 105,000 km^2 of forests in a total area of 179,800 km^2 (Li *et al.*, 2021). From 2005 to 2013, the carbon sequestered by vegetation in Guangdong ranged from 54.5 million tons to 53.2 million tons per year (Pei *et al.*, 2018). Forests were responsible for 93% of the carbon sequestered, although Guangdong lost over 4,000 km^2 of forest land then (Pei *et al.*, 2018). There is a higher density of forest cover in northern Guangdong compared to southern Guangdong (Li *et al.*, 2021). The average increase in forest carbon storage per year from 1979 to 2012 was estimated at 4.57 million tons per year, while the carbon storage density in 2012 was 26.77 tons of carbon per hectare (Li *et al.*, 2021). This density is lower than the average carbon density of forests found in China (38.05 tons) and much lower than the forest density in Central Africa (116.9 tons) and Central America (90.4 tons), primarily due to differences in forest types, origin, and management intensity (Li *et al.*, 2021). The low density is attributed mainly to the young age of the forests located in Guangdong (primarily young and middle-aged) and minimal incentive for local management (Li *et al.*, 2021).

4.4. *Infrastructure of carbon markets*

Carbon market is a widely adopted policy instrument for carbon neutrality. Guangdong launched a pilot emissions trading scheme (ETS) in 2013, and the full-scale Chinese ETS system was established in 2021. This smaller-scale system, as of 2016, encompassed 70% of the province's emissions, with 245 participants as of 2020 (ICAP, 2020), for a total cap of 465 million tons in 2020 using top-down and bottom-up methods. The primary goal is to decrease carbon intensity in line with the country's overall emission goals (Zeng *et al.*, 2021). The ETS system has affected secondary industries since 2013, with emissions increasing by only 3.1% from 2013 to 2020 and a corresponding fall of 21.6% in emissions per unit of GDP. Absolute CO_2 emissions from electricity, cement, iron & steel, and petrochemicals saw reductions of 12.3% during 2013–2019 (Zeng *et al.*, 2021), although their emissions nearly doubled from 2005 to 2013 (Pei *et al.*, 2018). Price was very low, especially in 2021, being only 40.2 RMB per ton on average, while the EU ETS system's average price was 53.91 euros per ton and the new UK ETS system corresponded to 48.11 pounds per ton (ICAP, 2022). Penalties for failing to submit emissions or verification reports are also low, at a maximum of only 50,000 RMB, or just over US$7000, with slightly higher penalties for not matching stated emissions (ICAP, 2020).

The Guangdong ETS system as a form of reducing total emissions can be seen as not harsh enough because 95% of the power sector's allowances were freely allocated, while the ratios for the aviation sector and other sectors were 100% and 97%, respectively (The World Bank, 2021). The system helped raise US$1.7 billion for the government (The World Bank, 2021). The Chinese national ETS system now regulates over 2,200 companies within the power sector, accounting for 4 billion tons of CO_2, or 40% of the country's annual emissions (ICAP, 2021).

This was built off the back of seven local and provincial pilot ETS systems in the country, leading to only a slight increase in the number of regulated companies from 1,961 in 2019 to 2,225 in 2021 (ICAP, 2021). As of late 2021, the Chinese ETS system has been operating without an absolute cap on emissions and with prices per ton of carbon at a tenth of the price in the EU ETS system (Roldao, 2022). For Guangdong to meet its carbon emissions goals, it is likely necessary for it to move to a higher pricing structure similar to the EU's, along with an absolute cap on total emissions. The EU uses a declining annual cap on emissions, with plans to reduce emissions by 55% in 2030 from the 2005 level (European Commission, 2021). As a form of revenue, carbon markets and carbon taxes could be effective ways to increase related investments. As another simultaneously implemented market-based policy instrument, carbon taxes in Europe averaged 35.91 euros per ton, covering 34% of all emissions as of April 2021 (Elke, 2021). The coverage and tax levels vary widely. In Ukraine, the cost per ton of emissions was only 0.25 euros compared to 116 euros in Sweden, while the coverage is also drastically different, from 3% in Spain and Lativa to 71% in Ukraine (Elke, 2021). France raised nearly US$10 billion of revenue during 2014–2021, on top of the income from the EU ETS system (The World Bank, 2021). To avoid double counting the same emissions, any company within the EU ETS system is exempt from the carbon tax, which was frozen at 44.6 euros per ton in 2019 (The World Bank, 2021). The EU ETS system led to a revenue of 57 billion euros from 2012 to June 2020 for the EU, with 14 billion in 2019 alone and 7.9 billion in the first half of 2020 (European Commission, 2021). By law, at least 50% of the revenue from the EU ETS must go to energy projects, and real investment from this revenue is around 78% from 2013 to 2019 (European Commission, 2021).

Chapter 5

Challenges and Opportunities

1. Challenges

1.1. *Technical challenges*

The availability of low-value land and water surfaces is a significant technical challenge for the Greater Bay Area (GBA), including Hong Kong, Macau, and nine other developed municipalities in Guangdong Province. This could prevent them from relying heavily on locally generated electricity from renewables. Among the three scenarios explored in Chapter 4, the Renewables Scenario has the highest demand for land and water surfaces for installing solar PV and onshore and offshore wind turbines. Wind and solar energy densities are constrained by nature, requiring about two orders of magnitude more land than coal- or natural gas-fired power plants. For example, Hong Kong's Black Point Power Station occupies an area of about 40 hectares with 3,225 MW of natural gas-fired power capacity at present and more land reserved for future expansion. The land requirement is then less than 125 m^2/MW. Solar PV requires about 100 times more land per MW in Europe, while wind energy could demand significantly more if the land between wind turbines is not used for other purposes (Trondle, 2020). Given that wind turbines and solar PVs tend to have much lower capacity factors than natural-gas-fired capacity, the land requirements will be even more significant to generate enough wind and solar electricity. In 2019, Hong Kong's local plants generated 36,795 GWh of electricity (HK Census and Statistics Department, 2022). If adopting the land requirement for solar electricity in nearby Vietnam (7.18–8.16 m^2/MWh) (Sanseverino *et al.*, 2021), Hong Kong would need about 300 km^2 of land to install enough solar PVs and decarbonize local electricity generation, which is equivalent to the total built-up area within the city. Accordingly, local renewable energy can play a minor role at best and is never decisive in decarbonizing Hong Kong's local electricity generation. This is supported by Hong Kong's 2050 climate plans, which admit that even in the most optimistic cases, renewables will generate less than 15% of Hong Kong's total electricity supply (HKGOV, 2021a).

Furthermore, weather and climate are other crucial factors that affect the capacity and availability of wind turbines and solar PV, which will then influence the reliability and resilience of electric systems. For example, the annual sunshine hours in Guangdong vary significantly across space, seasons, and years. In 2021, the eastern region of Guangdong had 2,567 h of sunshine, whereas the northern region had only 1,898 h. In 2016, their sunshine hours were only 1,701 and 1,629, respectively (Guangdong Bureau of Statistics and National Bureau of Statistics, 2022). Battery-based energy storage is mainly used to balance electricity demand and supply within a day or a relatively short period, but not across seasons or even years. These temporal variations of wind and solar energy indicate that renewables alone can hardly provide a reliable electricity supply.

1.2. *Financial challenges*

The financial challenges are twofold: the much greater capital requirements for carbon-neutral infrastructures and the stranded assets of fossil-fuel infrastructures. Massive investment is required in the coming decades for the region to approach carbon neutrality. Although we apply conservative estimates of future electricity consumption growth and energy infrastructures' capital costs, our results suggest that over the 35 years of 2026–2060, US$611.5 billion, US$572.4 billion, and US$480.2 billion would be demanded in the Renewables, CCS, and Nuclear Scenarios, respectively. Besides the related operation & maintenance costs that they incur annually throughout their lifetimes, the choice of different pathways to approach carbon neutrality in the region could sway funding well over US$1 trillion. Strategic mistakes could create stranded assets from these new investments.

Stranded assets could be a severe problem when energy infrastructures must be phased out prematurely or their utilization should be scaled down. In 2020, the region's energy infrastructure asset values were 69.9% in fossil fuels. They could incur potential losses, especially in the Renewables and Nuclear Scenarios. Even though they may continue generating electricity until reaching their designed lifetimes, their capacity factors may be forced to dwindle over the years, thus throwing their profitability into question.

Estimates within China of the cost of decarbonization to reach net zero in 2060 are around US$20 trillion by 2050 (Nedopil and De Boer, 2020). Under assumptions for meeting the Paris Agreement targets (2°C scenario), just under 130 trillion RMB would be required, with around 100 trillion RMB in the energy sector. In contrast, a 1.5°C-compatible scenario would require an investment of over 170 trillion RMB (Nedopil and De Boer,

2020). World Bank predictions vary slightly, showing expected investments of US$17 trillion for peak emissions in 2030 and net zero in 2060 (Ylhe, 2022).

1.3. *Social challenges*

The fossil-fuel industry still employs many people within the GBA, Hong Kong, Macau, and other Guangdong municipalities. Unemployment could be a severe challenge in the Renewables and Nuclear Scenarios. In Hong Kong alone, for example, CLP and HKEI employ over 6,400 people within the power sector. While most employees work within their transmission & distribution branches, the three primary power plants employ many people, although the exact distribution is unavailable (CLP, 2021; HKEI, 2021). In September 2022, 11,763 people were employed in the energy, gas, or waste management industries (Census and Statistics Department, 2022), in comparison to the total Hong Kong employment figure of around 3.7–3.8 million people (Census and Statistics Department, 2023). For Guangdong, coal power plants alone employed over 22,000 people, while the total employment in China was around 469,000 people (Cui *et al.*, 2020). Data from the Guangdong Yearbook show total employment within the electric gas and water industry in 2019 at 366,300 people out of the total employment of 71.5 million people (Guangdong Bureau of Statistics and National Bureau of Statistics, 2020). Although these are not very large numbers compared to the total employment figures, the governments should find ways to transition these people toward more sustainable energy jobs. For Macau, since most energy is imported, the number of people employed directly in energy production is limited. CEM employed 712 people at the end of 2019, an increase of only two from 2010 (CEM, 2021a). Total employment in Macau in the same year was 394,000 (DSEC, 2022).

Although the operation and maintenance of carbon-neutral infrastructures may not create fewer jobs than the replaced fossil-fuel infrastructures, the required skills in those jobs are not easily transferable. Furthermore, their geographical distributions do not match each other. Jobs may shift from Hong Kong and the GBA to non-GBA Guangdong or even beyond to other western provinces due to the latter's abundance of land and shorelines with much less intensive economic activities and population. Newly created jobs toward carbon neutrality may not have readily available employees with well-matched skills. Training should be planned to equip potential employees with professional knowledge and skills.

1.4. *Scalability challenges*

Scaling up the technologies and energy systems required for this change could be a monumental challenge. In the past two decades, substantial progress has been made on solar PV, onshore and offshore wind, and batteries. The technologies have been highly commercialized even to the point of the phase-out of subsidies. During 2021–2025, the region is projected to add 38.1 GW of solar PV and onshore and offshore wind capacities, or 7.6 GW/year. In comparison, the Renewables Scenario requires the pace to double to 14.4 GW/year, along with 2.2 GW of battery-based energy storage annually.

Nuclear energy is also being steadily developed, especially in China. However, due to serious concerns about nuclear safety, especially in the aftermath of the Fukushima nuclear accident, nuclear power plant installations tend to adopt a much longer process, from site planning to construction. This could become a significant constraint on scalability. Over the 35 years of 2026–2060, the Nuclear Scenario requires the addition of 155.2 GW of nuclear power plants, or 4,435 MW/year. In contrast, our data on in-pipeline projects account for 21.7 GW of capacity growth from 2025 to 2032, or 2,713 MW/year, the fastest pace ever in the region. The rate should be doubled to realize the Nuclear Scenario.

The CCS Scenario may have the biggest scalability challenge because CCS technologies have not been commercialized. Guangdong and even the entire China have not completed a single full-scale, whole-chain CCS project in fossil-fuel-fired power plants. Major CO_2 transport infrastructures, such as pipelines and ships, do not exist. CO_2 storage sites that can store about half a billion tons of CO_2 annually have not been verified and developed.

1.5. *Institutional challenges*

Infrastructure projects' planning, design, approval, and construction often take many years or even decades. Many factors could slow down the process. Due to China's industrial prowess, the manufacturing of essential and technologically mature equipment may not be the narrowest bottleneck. At the same time, institutional barriers may raise different challenges to implementing the scenarios.

Past experiences suggest that Hong Kong will not accept a nuclear power plant within its territory. At the same time, Hong Kong and the GBA will not have sufficient land and water surface for solar PV and wind turbines to fulfill their electricity demands. These two scenarios in Hong Kong's context would indicate that its electricity demand will not be met by electricity generation within the geographical territory of Hong Kong. It should either abandon/scale down the current electricity independence and rely on imported electricity from China Southern Power Grid or expand the dedicated transmission lines with power plants in Guangdong Province. The Hong Kong electricity market is highly regulated, with two vertical monopolies featuring geographically separate zones of control and limited interconnection. China Southern Power Grid operates a regulated electricity market, allowing for electricity transmission and generation to be managed by independent companies. The two scenarios come with severe challenges from the incompatibility of current electricity markets in Hong Kong and the GBA. The solution may require Hong Kong to deregulate its monopoly market and encourage competition. The current scheme of control agreement (SCA) was signed with CLP and HKE in 2017 for a 15-year term from 2019 to 2033 (Hong Kong Electric, 2022). The scheme controls the permitted return for the power companies at 8% of the total value of their average fixed assets for that year. It is also used to encourage increased energy efficiency and renewables usage. The SCA disallows price gorging, as fuel costs and expected tariffs (costs) passed on to the customer are highly regulated and audited by the government to maintain lower associated costs (Hong Kong Electric, 2022). As such, Hong Kong's electricity price of HK$1.09–1.22/kWh is similar to that of Shenzhen (HK$0.88/kWh) or Seoul (HK$0.96/kWh) but much lower than comparable cities such as London (HK$2.7/kWh) or New York (HK$2.7/kWh) (Hong Kong Electric, 2022).

The CCS Scenario could maintain the status quo of the current fossil-fuel-dominated energy system; however, it has the most significant technological and scalability challenges. Like wind, solar, and battery technologies, commercializing the currently immature CCS technologies mandates public support through various subsidies. However, the expected high subsidy budget and corresponding financial risks could lead to significant institutional challenges. The usual private–public partnership model may not be directly replicable.

CO_2 emission trading is an essential climate policy for incentivizing investment in carbon-neutral energy infrastructures. Although China has established a national emission trading market, its institutional design and practical implementation have not fulfilled their intended purposes. The carbon price in the Chinese ETS market is about one order of magnitude cheaper than in the EU ETS system (EU, 2021). Although it covers 10% of global emissions, the Chinese market is estimated to have a transaction volume of only US$1.4 billion in 202 (Climate Action Tracker, 2023).

2. Opportunities

Challenges and opportunities generally go hand in hand. If Hong Kong and the GBA could actively and successfully meet the challenges of carbon neutrality, the region would be transformed to embrace a more sustainable future.

2.1. *Economic opportunities*

Hong Kong and the GBA barely have any fossil-fuel resource endowments. Reducing their imports under carbon neutrality could thus leave a more significant chunk of the expenditures within the region. The transformation of the energy sector will lead to significant economic changes and investment in emerging technologies. The trillion-dollar-level expenditure for the region to achieve carbon neutrality also corresponds to a trillion-dollar-level opportunity for related industries and employment: renewables, energy efficiency, and fossil fuels are estimated to create 7.49, 7.72, and 2.65 jobs per million US dollars spent, respectively (ADB, 2021).

Hong Kong and, especially, the GBA aim to build and strengthen innovation and technology industries. Carbon neutrality creates tremendous demand for expanding green infrastructure, phasing out fossil-fuel infrastructure, developing new technologies, and adapting technologies to local contexts. The Ukraine–Russia war accelerated the gains associated with energy efficiency on a global scale. Predictions from 2022 anticipated a 2% increase in global energy efficiency in that year, four times as much as in 2021 (0.5%) (IEA, 2022a). These helped achieve a 1% drop in energy consumption in 2022 compared to 2021 (IEA, 2022a). However, 2021 saw the most significant increase in year-on-year energy usage in 50 years, at 5% (IEA, 2022a). As a part of infrastructure spending (urban retrofits, public transport, and EV support), spending on energy efficiency crossed US$1 trillion during 2020–2023, with over US$700 billion from private spending (IEA, 2022a). Therefore, R&D and investment in improving energy efficiency are areas on which Hong Kong can focus, as two-thirds of sustainability funding already goes to this area (IEA, 2022a).

The geographical distribution of these new carbon-neutral infrastructures is leaning toward much poorer non-GBA Guangdong and beyond, especially in the Renewables Scenario. Carbon neutrality could thus greatly benefit those municipalities and provinces by boosting their GDP. The economic geography of the energy economy will also change dramatically. Intra- and interprovincial disparities could thus be alleviated.

2.2. *Financial and professional service opportunities*

Carbon neutrality creates an excellent opportunity for the financial and professional services sectors in Hong Kong and Guangdong Province. Professional services could offer significantly higher salaries than average employment. Hong Kong and the GBA are high grounds for professional services, with a concentration of highly skilled professionals from all over China and the world.

As a global financial center, Hong Kong is well positioned in climate finance. Reworking and decarbonizing the entire energy sector will allow for extensive market opportunities in both the public and private sectors. The World Bank highlighted how, from 2020 to 2030, nearly US$25 trillion of market opportunities for green building alone would appear for emerging markets' cities (The World Bank, 2020). According to Bloomberg, the environmental, social, and governance (ESG) market had an expected total assets under management (AUM) of US$53 trillion by 2023 (Bloomberg, 2021). Paired with the increase in green financing from the ESG debt market (US$2.2 trillion in 2020), especially with green bonds, this could pave the path for more investments in green tech (Bloomberg, 2021). The ESG debt market alone could grow to US$11 trillion by 2025, and green bonds have already seen US$2 trillion in cumulative volume by the end of 2020 (Bloomberg, 2021). According to the Climate Policy Initiative (CPI), investments in climate financing reached US$632 billion in 2019/2020; however, to reach the IPCC's goals, financing would need to increase to US$4 trillion per year by 2030 (CPI, 2021). Currently, funding is nearly equal between private and public sources, with 51% being public financing, while the vast majority (over 90%) of funds are used for climate mitigation and only 7% for climate adaptation (CPI, 2021). This lack of adaptation funding is a serious issue, with the UN estimating that developing nations will need US$155–330 billion annually by 2030 for their adaptation needs (CPI, 2021).

CO_2 emission trading is a heavily financialized service. Guangdong and Shenzhen have had two CO_2 emission trading pilot schemes since 2013 (ICAP, 2020), and China established a national market in 2021. The Hong Kong Exchange service announced in October 2022 the creation of a carbon marketplace called Core Climate (HKEX, 2022). This current market still only allows the dealing of voluntary credits on the market and is not part of the China ETS system, although the possibilities are being explored (HKEX, 2022). A memorandum of understanding with the Guangzhou Future Exchange was signed to explore greater sustainability-related cooperation (HKGOV, 2021a).

The GBA and Guangdong may benefit from an integrated carbon market. By the end of 2019, Guangdong's ETS system covered 60% of the province's emissions, covering the power, steel, cement, and petrochemical industries, paper making, and domestic aviation (ICAP, 2020). Furthermore, the power sector was integrated into the national ETS system in 2021, allowing for a broader scope for trade (International Carbon Action Partnership, 2021). For smaller systems and integration, the use of the Guangdong ETS system coincided with a 12.3% reduction in Scope 2 emissions from the steel, cement, electricity, and petrochemical industries from 2013 to 2019 (Zeng *et al.*, 2021). Although not entirely due to the application of the ETS system, cost pressures from this system did lead to 66 enterprises either stopping or reducing operations to meet the cap set under it, with these enterprises producing nearly 15 million tons of emissions before the ETS system (Zeng *et al.*, 2021). Integrating the Hong Kong energy system into both the national and provincial level systems could be one way of reducing the overall emissions associated with power production within the city while also bypassing some of the rigidity of the SCA (HKGOV, 2019b; ICAP, 2020; International Carbon Action Partnership, 2021; Zeng *et al.*, 2021). More industries that are being considered for being included in the Guangdong ETS system (data centers, ceramics, textiles, transportation, and construction) could further deepen the value of the ETS system (Zeng *et al.*, 2021). The construction and transportation sectors are highlighted as areas that could help bring Macau and Hong Kong into the ETS system, although challenges exist (Zeng *et al.*, 2021).

ESG is another rapidly expanding financial and professional services market. As part of new regulations in 2019, all Hong Kong-listed companies on the stock exchange now have an ESG reporting requirement as of 2021 (Conventus Law, 2019). The stock exchange mandates that companies hold new climate change provisions and environmental target disclosures (Conventus Law, 2019). Green bonds (fixed-income instruments used to finance new and existing green projects) have also seen a significant rise in popularity within the financial sector (HKIMR, 2020). As the market is new, the classification of what bonds are considered green does change depending on location and when the bond was certified (HKIMR, 2020). For example, China had classified clean coal as a technology under the 2015 definition of a green bond, although this was removed in the 2020 revision (HKIMR, 2020). Global green and other sustainability-aligned bonds have gone from under US$200 billion in issuance in 2015 to over US$1 trillion in 2021 (Harrison and Muething, 2022). Green bonds alone saw a 75% increase in issuance between 2020 and 2021, from under US$300 billion to over US$520 billion in new issuances, while sustainability-linked bonds increased by over 900%, from US$11.4 billion to US$118.8 billion (Harrison and Muething, 2022). According to S&P Global, every US$1 million investment in green bonds could lead to emission reductions of up to 3,000 tons (Helfre and Depetiteville, 2022). For a location such as Hong Kong, where the primary energy usage is from buildings, the avoided emissions per million US dollars will be significantly lower (Helfre and Depetiteville, 2022). That said, Hong Kong has been a repeat green bond issuer, issuing US$7.2 billion in sovereign green bonds by year-end 2021 (Harrison and Muething, 2022). Total issuance in Hong Kong had reached US$26 billion by the end of 2019, dominated primarily by mainland China entities (HKIMR, 2020). Hong Kong also created a green bond guide in November 2020 (HKIMR, 2020). As a financial instrument, the Hong Kong government is bullish on green bonds, doubling borrowing from HK$100 to HK$200 billion (HKGOV, 2021a). Further integrations between the Hong Kong and Guangdong governments are considered likely due to the growth of green bond issuance from China. The Hong Kong Stock Exchange aims for Hong Kong to become a hub for Asia's sustainable finance market (HKGOV,

2021b). Further client data from the larger Hong Kong banks show significant interest in ESG management and ESG products in general (HKGOV, 2021b). Within Greater China, this does leave an area for collaboration, especially since, as of 2019, 85% of CSI 300 companies had released official ESG disclosures, up from 54% in 2013 (Tan, 2020). However, it is crucial to note that although disclosure rates have been rising, the quality of these disclosures is lackluster, with only 12% of disclosures being audited, up by only 1% from 2013 (Tan, 2020).

Chapter 6
Conclusion and Implications

1. Conclusion

Hong Kong, the Greater Bay Area (GBA), and Guangdong Province have trillion-dollar stakes in energy infrastructures for achieving carbon neutrality. The role of decarbonization is already a pressing issue for the Guangdong provincial government. It is becoming one of the most highly prioritized governmental affairs in Chinese politics and economic development. Several major conclusions can be reached through our analysis:

- **A variety of energy infrastructures are needed to support each other.**

Three distinct scenarios are worked out for Guangdong Province and Hong Kong to approach carbon neutrality. Power plants that have been planned will be constructed as scheduled. To meet all additional electricity demands in future years (such as due to electric vehicles and the usual growth of energy consumption), the three scenarios assume the following: exclusive reliance on wind and solar with energy storage (Renewables Scenario), coal and natural gas-fired power plants with CO_2 capture and storage (CCS) (CCS Scenario), and nuclear power with energy storage

(Nuclear Scenario). More scenarios could be constructed by combining the three pathways.

The three scenarios distinguish the importance of offshore wind, nuclear power, CCS, and energy storage for the region's local electricity generation. For any remaining gap between local electricity generation and consumption, electricity should be imported from other provinces, which is already significant and could occupy an even greater proportion. Remaining non-electrified CO_2 emissions can be decarbonized by biofuel and hydrogen or offset by nature-based carbon sinks and credits purchased in carbon markets.

None of the three scenarios is without flaws, and all have strengths, as summarized in Table 1. The Renewables Scenario requires a vast amount of land and water surfaces. The electricity

Table 1. Challenges and opportunities of the three scenarios.

	Renewables Scenario	CCS Scenario	Nuclear Scenario
Dispatchability	☹	😍	☹
Reliability and resilience	☹	😍	😍
CO_2 emissions	😍	😐	😍
Land and water surface requirement	☹	😍	😍
Rebalancing economic geography	😍	☹	😐
Technological readiness for rapid deployment	😍	☹	😐

system could become less reliable and resilient, especially across seasons and years, because battery-based energy storage will mainly help balance electricity supply and demand within a day or a short period. It can effectively decarbonize electricity and shift economic opportunities to poorer municipalities outside the GBA. The CCS Scenario still has much residual CO_2 emissions, although the emission factor may be reduced by one order of magnitude. It may further concentrate the energy infrastructures in the more developed GBA, which does not help address Guangdong Province's severe internal regional disparity. They have the best dispatchability to enable a reliable and resilient electric system, which becomes increasingly important with further economic development. The Nuclear Scenario can provide a highly reliable electricity supply across seasons and years; however, nuclear power plants lack adequate dispatchability, as shown by their currently almost saturated capacity factors. Significant energy storage capacities will be required to rebalance the diurnal cycles of electricity demand.

As a result, the region should not rely on any of the three scenarios alone. Their major technologies – renewables, especially offshore wind, nuclear energy, CCS, and energy storage – should all become critical components in future energy systems. This conclusion should also be valid for many other coastal regions/countries with land availability constraints due to high population and economic densities, such as Taiwan, the Yangtse River Delta, Japan, and many European countries at higher latitudes. However, the proportions of these different energy infrastructures must be adjusted according to their contexts.

Carbon neutrality requires a fundamental transformation of energy-related infrastructures within one human generation, while time is an increasingly scarce resource to meet the imminent deadline. The most important way to minimize stranded assets is to rapidly reduce new investments in fossil-fuel infrastructures if they are not equipped with CCS. Investments in renewables, nuclear, and other new types of energy infrastructure will be required at a substantial pace, with additional capital investment of a few hundred billion US dollars even under our highly conservative estimates. Annual investments are peaking at over US$30 billion per year as we reach 2060. The region must start planning those energy infrastructures as time passes to meet the carbon neutrality targets in 2060, 2050, or earlier.

Nevertheless, challenges and opportunities go hand in hand as two sides of the same coin. The metropolitans' energy infrastructures will be fundamentally transformed as they approach carbon neutrality. However, carbon neutrality may target Scope 1, 2, or 3 emissions with significantly diverging implications. Scope 1 emissions of a region refer to those CO_2 molecules that enter the atmosphere geographically within its boundaries. Scope 2 mainly accounts for embedded emissions in the electricity trade. Scope 3 emissions are life-cycle emissions of goods and services consumed in the region, while international and inter-regional trade indicate that the emissions may be widespread globally. Furthermore, metropolitan regions, such as Hong Kong, Macau, and the GBA, may find purchasing emission credits in emission trading markets cheaper than local mitigation. They would also forgo great opportunities if their carbon neutrality commitments were only half-hearted, i.e., relying on purchased emission credits to offset emissions and circumventing those challenges. Furthermore, when China and other major economies inch closer to carbon neutrality in the middle of the century, the supply of CO_2 emission credits would be deeply constrained, but the demand could surge. As a result, the costs of relying on purchased emission credits will be much higher than today. Accordingly, this option may delay painful actions for carbon neutrality but not avoid them. If so, other economies may have already harvested the opportunities, while latecomers may face fierce competition with many incumbents.

- **Stranded assets of existing/planned fossil-fuel infrastructures will not be China's primary obstacle to reaching carbon neutrality in 2060, but any future addition of these infrastructures could face barriers.**

Fossil-fuel infrastructures currently account for the lion's share of China's energy infrastructure asset values. The prospect of potentially stranded assets has become a severe concern about China's carbon neutrality commitment for 2060. Our research concludes that currently existing or planned fossil-fuel infrastructures will not be a significant obstacle from the perspective of stranded assets, even if China must phase all of them out by 2060. We quantify assets as an infrastructure's book value. Its current book value is the maximum amount of stranded assets an infrastructure can incur. It is assumed to depreciate linearly within an infrastructure's expected lifetime and will be written off when the remaining lifetime becomes zero. This study does not recognize lost revenues in the future or the monetary valuation in stock markets as assets, which were used in other studies. The different definitions of assets could lead to different conclusions on the sizes and impacts of stranded assets.

This study focuses on Guangdong, a province adjacent to Hong Kong that is responsible for over 10% of China's GDP and energy consumption. The existing and planned electricity generation capacities geographically located in the province exceed 200 GW, while much of the electricity is imported from other provinces. Accordingly, the primary conclusion above should apply to all of China without significant deviations.

Most asset values of current fossil-fuel infrastructures, either already in operation or planned to be, will be depreciated or written off before 2060. As a result, their potentially stranded assets will not take up a significant share of the total asset values of energy infrastructures in 2060. The depreciation will take place gradually over a few decades, rather than being a one-time event, which creates shocks. The financial penalty for China (or

Guangdong Province in this study) would not be unbearable, even if they must be written off. Nevertheless, their earlier phase-out or phase-down will increase the proportion of stranded assets. Hong Kong's replacement of coal-fired power plants with natural-gas-fired ones has a profound lock-in effect of incurring significant write-offs before retirement. Natural gas, although significantly cleaner than coal used in sub-critical power plants, such as Castle Peak, still leads to about half of CO_2 emissions per kWh (Environmental Protection Department, 2022). The increase in natural gas infrastructures through the creation of both a new receiving terminal near Soko's islands and new natural gas turbines at both Lamma and Black Point shows a lock-in of emissions, making it significantly more challenging to meet the 2050 goals, even if all turbines are shut down once they hit their 30-year lifetime. Instead of being a bridge to carbon neutrality, the ongoing energy transition toward natural gas in Hong Kong and the GBA could significantly become an obstacle to realizing its carbon-neutral future if the timeline is advanced.

A great majority of electricity demand in 2060 will be met by power plants that do not exist or have not been planned by now. Future choices, such as between the Renewables and CCS Scenarios, are much more critical. In other words, today's energy infrastructures will not significantly influence China's energy mix in 2060. New investments in the coming decades will mainly define the problem of stranded assets. Although Hong Kong and the GBA are rapidly transitioning from coal to natural gas in local electricity generation, new investment in fossil-fuel infrastructures without CCS should be minimized. This leads to a change in the investment location away from the GBA for energy infrastructure and closer to non-GBA areas with cheaper land. Light-asset infrastructures are better at serving as a bridge toward carbon neutrality. As renewables have reached grid parity with fossil-fuel-fired electricity, investment in renewables tends to be much more resilient in a competitive electricity market.

Continuous investment in natural gas infrastructures is not a promising bridge to carbon neutrality.

Fossil-fuel-fired electricity tends to be significantly less asset/capital-intensive than that of renewables and nuclear energy, as measured in proportions of the levelized costs of electricity. Fuel costs and other operation/maintenance costs are often the most significant proportion of, or dominate, the levelized costs of fossil-fuel-generated electricity (even with CCS due to its heavy energy penalty). At the same time, wind turbines and solar panels tend to have much lower capacity factors with shorter lifetimes (thus requiring more capacities to generate the same amount of electricity). When the two sources of electricity have similar costs, renewable energy infrastructures (wind turbines, solar panels, and battery storage) and nuclear power require much higher asset values to generate one kWh of electricity.

Fossil-fuel infrastructures are disproportionally located in relatively wealthier regions (Guangdong Province and the GBA). In comparison, renewable infrastructures are more concentrated in relatively poorer regions (even within Guangdong Province), partly driven by their land/coastal water values. The write-off of potentially stranded fossil-fuel assets in 2060 will be mainly in regions with deeper pockets. The Renewables Scenario will contribute to the alleviation of regional economic disparities.

- **Hong Kong and the GBA cannot achieve carbon neutrality alone and cooperation is necessary.**

Hong Kong, Macau, and other metropolitans in the GBA do not have the space required to be energy independent. They do not have the required land or coastal waters for large-scale renewable energy development in the Renewables Scenario. Their space should not be mainly used for this purpose either, because their current economic activities per km^2 can produce much greater outputs than wind turbines and solar PV. In the case of Hong

Kong, even with both currently planned large-scale offshore wind farms, the Hong Kong government expects only 7.5–10% of Hong Kong's energy production will be from renewables (HKGOV, 2021a). Looking at the current production amount, however, it seems unlikely that Hong Kong will meet these goals, as only 0.3% of Hong Kong's electricity is currently being produced via renewable energy (Electrical and Mechanical Services Department, 2021).

Furthermore, metropolitans do not have a high willingness to host enough nuclear power plants within their territories, which constrains their independent application of the Nuclear Scenario. The CCS Scenario mandates the safe storage of a large amount of CO_2 (a dangerous asphyxiant gas that causes hypercapnia if leaked suddenly in significant quantities), while metropolitans do not have suitable and large enough storage sites. As a result, they must work with others to achieve carbon neutrality.

2. Implications on Politics vs. Humanity

Climate change and human responses are heavily shaping the world today and will continue to do so in the future. The two major topics at the annual United Nations Climate Change Conference, humanity and politics, witness both synergies and conflicts in this global arena of high stakes. Uncertainties still loom over which side can triumph, although humanity must prevail for homo sapiens to have a bright future on this lonely planet.

Humanity mandates global cooperation, including healthy competition that could better facilitate its goals. The world in the past two decades, in comparison to the preceding two decades, has experienced very different trajectories of CO_2 emissions (Figure 1). There has never been such a long streak of peaceful

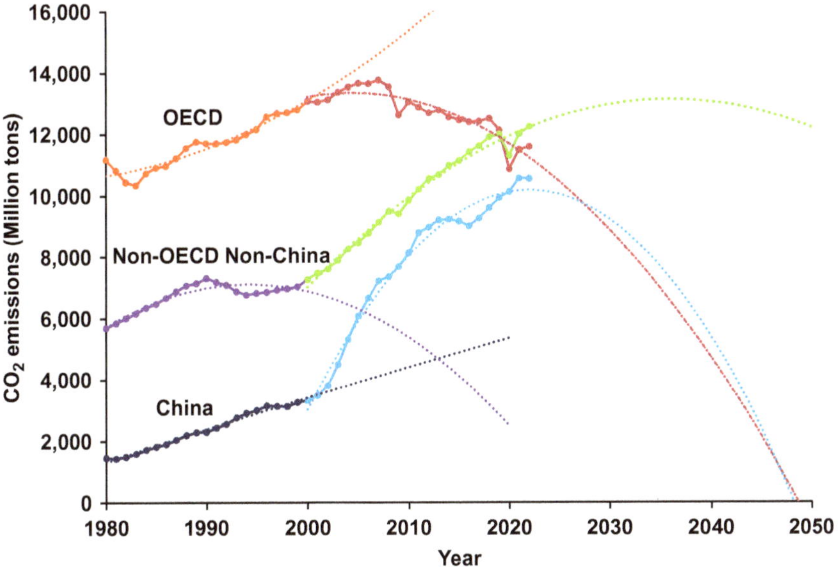

Figure 1. Projections of CO_2 emissions based on actual emissions during 1980–2000 and 2000–2022 (Energy Institute, 2023).

time among or within all major economies. The advancement of global cooperation achieved a significantly higher degree of globalization and yielded many dividends to substantially elevate their citizens' living standards. Prosperity had a significant side effect of enhancing energy consumption and thus CO_2 emissions from fossil fuels, but it also resulted in their fundamental solutions. Wind turbines, solar PVs, and batteries are the most successful climate mitigation technologies developed in the past two decades. The significant advancement of renewable energy, batteries, and electric vehicles, as critical enablers for carbon neutrality commitments, was the product of practical global cooperation. Globalization was crucial for the world to enjoy economic growth and solutions to climate change. In the environmental Kuznets curve framework, peak emissions may have come earlier and lower due to global cooperation (Figure 1).

It also promotes innovation and the deployment of new technologies. While the former effect dominated the 2000s, especially in China and non-OECD non-China countries, to raise global emissions, the latter effect is increasingly essential for accelerating the downward trends.

As a sign of progress, they have witnessed paradigm shifts for all three groups of countries over the past two decades (Figure 1). OECD countries have had their CO_2 emissions peak, while their earlier trend steadily increased. China transitioned from a gradual and nearly unidirectional climb to a trajectory with a rapid initial skyrocketing lift, followed by a period of plateauing, to a probably imminent peaking and then a swift decline. Non-OECD, non-China countries experienced an increase in emissions rather than a decrease, which is reflected in their faster economic and energy consumption growth and better living standards. The world witnessed the entry into force of the Kyoto Proposal in 2005, the Paris Agreement in 2016, and the carbon neutrality pledges by all significant emitters during 2019–2021.

If the paradigm shifts indicated that humanity prevailed over politics, the recent reorientation to politics may slow down the changes. China's CO_2 emissions have grown faster since the breakout of the US–China trade war in 2018, despite the impacts of the devastating COVID-19 pandemic and skyrocketing capacities of wind and solar. They were 13.8% higher in 2022 than in 2017, while the growth rate was only 3.3% from 2012 to 2017. Suppose the winning streak of humanity could continue in the next two decades. In that case, we should witness the further bend-down of OECD and Chinese CO_2 emissions and hopefully accelerate the downward trend in emissions of those non-OECD, non-China countries.

However, geopolitical conflicts, national security, decoupling/de-risking, export/import restrictions, or even complete bans, sanctions, and retaliation are increasingly inching into the mainstreams of nations. The reverse globalization trends and segregated supply chains may slow future innovation and deployment. Countries' fates could change dramatically in the megatrend of climate change and responses. While the United States has been transformed in the past 15 years from the most significant energy importer to a major energy exporter, China's oil and natural gas imports now top the world, raising serious energy security concerns. The carbon neutrality pledge that mandates substantial reductions in fossil fuel consumption offers tremendous opportunities for developing wind, solar, energy storage, electric vehicles, other technologies, and related services. Many fossil-fuel value chains depend on immobile resource endowments, but much smaller shares of the carbon-neutrality technologies' value chains are geographically fixed. This leads to serious trade disputes when countries perceive their economic activities and employment disproportionally flowing out. These economic concerns have been increasingly politicized, even to the level of hard-to-compromise national security.

Suppose the current rebalancing between humanity and politics on climate change continues. In that case, we may especially witness the slow-down of the energy transition and, thus, the weakening of CO_2-reduction effects in the trajectories. This political trap may be even worse, as the world has higher CO_2 emissions than at the turn of the century. OCED countries and China may see a slower reduction of CO_2 emissions, and the rest of the Global South may be stuck at high emission levels. Then, the world may lose its hard-earned momentum to reduce emissions. The Global South could decide the future of climate change. Still, the souring geopolitics between the United States and China, or an economic cold war, could jeopardize their future trajectory.

References

2022 Foundation (2019) GBA Background Document ... The Future of the Guangdong-Hong Kong-Macao Greater Bay Area (January 2019), pp. 1–221.

ACT (2018) Acron Project. *ACT.*

ADB (2021) *Implementing A Green Recovery in Southeast Asia.* ADB Briefs.

Andersson, J. and Grönkvist, S. (2019) Large-scale storage of hydrogen. *International Journal of Hydrogen Energy*, 44(23), 11901–11919. doi: 10.1016/j.ijhydene.2019.03.063.

Argus (2020) EV sales in 2020 to match 2019 despite Covid-19: IEA. *Argus.*

Baiyu, G. (2021) Does coal still have a role in China's decarbonising power market. *China Dialogue.*

Bermudez, J. M., Hasegawa, T., and Bennett, S. (2020) Hydrogen.

BHP (2020) Pathways to decarbonisation episode two: Steelmaking technology. *BHP.*

Bloomberg (2021) ESG assets may hit $53 trillion by 2025, a third of global AUM.

Bloomberg (2022) Gas & LNG - 10 predictions for 2022.

Brasington, L. (2020) Hydrogen in China. *Clean Tech Group.*

Brown, T., Anderson, B. G., and Schlachtberger, D. P. (2018) The role of hydro power, storage and transmission in the decarbonization of the Chinese power system. *Ground AI.*

Business Times (2020) Sinopec merges two Guangdong-based refineries lifts more US Oil. *Business Times.*

BYD (2021) BYD 2021 Interim Report.

Carbon Brief (2020) Mapped: The world's coal power plants.

Carbon Tracker (2020) Coal developers risk $600 billion as renewables outcompete worldwide. https://carbontracker.org/coal-developers-risk-600-billion-as-renewables-outcompete-worldwide/.

Catsaros, O. (2023) Global low-carbon energy technology investment surges past $1 trillion for the first time. *Bloomberg*.

CEM (2020a) CEM Annual Report 2019_small.

CEM (2020b) CEM Sustainability Report 2019.

CEM (2021a) CEM Annual Report 2020.pdf.

CEM (2021b) CEM business overview. *CEM.*

Census and Statistics Department (2022) Quarterly Report of Employment and Vacancies Statistics (September).

Census and Statistics Department (2023) List of All Statistical Products, Hong Kong SAR Government. https://www.censtatd.gov.hk/en/page_1226.html.

CGN (2020a) 2019 Annual Report.

CGN (2020b) CGN 2020 Interim Report.

CGN (2020c) Nuclear power. *CGN.*

Charlie, C. (2019) China is bankrolling green energy projects around the world. *Time*.

Cheng, C., *et al.* (2018) Reform and renewables in China: The architecture of Yunnan's hydropower dominated electricity market. *Renewable and Sustainable Energy Reviews*, 94 (June), 682–693. doi: 10.1016/j.rser.2018.06.033.

China Daily Global (2020) Energy in China's new era.

China Electricity Council (2020) Statistics of the power sector from January to September 2020. *China Electricity Council.*

China Electricity Council (2022) Annual Report of National Power Sector - 2021. https://www.cec.org.cn/upload/1/editor/1642758964482.pdf.

China Energy Portal (2020) 2020 Q2 PV installations utility and distributed by province. *China Energy Portal.*

China Southern Power Grid (2021) *Fortune*.

China State Grid (2019) Corporate profile.

Chung (2020) A coal-free future for Hong Kong. *City U.*

Circular Carbon Economy (2020a) Remove: Carbon capture and storage.

Circular Carbon Economy (2020b) Carbon capture and storage.

Climate Action Tracker (2021a) No Glasgow's 2030 credibility gap: Net zero's lip service to climate action.

Climate Action Tracker (2021b) USA.

Climate Action Tracker (2023) China. *Climate Action Tracker.*

CLP (2016) Guangdong Daya Bay nuclear power station. *CLP.*

CLP (2019) CLP's climate vision 2050. A more sustainable world.

CLP (2020a) CLP Annual Report 2019.

CLP (2020b) CLP information kit.

CLP (2021) Annual Report 2020.

CNOOC (2019) CNOOC 2019 Annual Report, CNOOC.

CNOOC (2020) CNOOC 2019 Environmental Social and Governance Report.

CNPC (2020) CNPC Annual Report 2019.

Conventus Law (2019) New ESG reporting requirements for Hong Kong listed companies from July 2020. *Conventus Law.*

Cornot-gandolphe, S. (2014) China's coal market: Can Beijing Tame 'King Coal' ?

CPI (2021) Climate financing. *Climate Policy Initiative.*

CSG (2015) Extra high voltage power transmission company. *China Southern Power Grid.*

CSG (2019) China Southren Power Grid 2018 Corporate Social Responsibility Report.

CSG (2020a) The Wudongde project, one of the key west-east electricity transmission projects has been placed into operation. *CSG.*

CSG (2020b) Volume reaches new high for 9th consecutive year. *China Southern Power Grid.*

Cui, S., Russell, P., and Zhao, J. (2018) Munich personal RePEc archive restructuring the Chinese freight railway: Two scenarios (88407).

Cui, R., *et al.* (2020) A high ambition coal phaseout in China: Feasible strategies through a comprehensive plant-by-plant assessment.

Daiqing, Z., *et al.* (2013) Analysis-of-CO2-emission-in-guangdong-province-china.pdf.

Darrell, P. (2020) DOE supports CCUS retrofit for San Juan coal plant. *POWER*.

DSEC (2022) Labour force, employed population and unemployed population by sex. *Macau Government*.

Duggal, H. (2021) Infographic: What has your country pledged at COP26? *Aljazeera*.

Dunn, C. and Barden, J. (2018) More than 30% of global maritime crude oil trade moves through the South China Sea. *EIA*.

Egan, M. (2020) China is storing an epic amount of oil at sea. Here's why. *CNN*.

EIA (2020) Country analysis executive summary: China.

EIA (2023) Units and calculators explained. *EIA*.

Electrical and Mechanical Services Department (2021) Hong Kong energy end-use data 2019.

Elke, A. (2021) Carbon taxes in Europe. *Tax Foundation*.

EMSD (2015) Natural gas. *EMSD*.

Energy Institute (2023) Statistical review of world energy. https://www.energyinst.org/statistical-review.

Energy Storage and Transportation (2021) Energy storage.

Environment Protection Department of Hong Kong (2020) Promotion of electric vehicles in Hong Kong. *Environment Protection Department of Hong Kong*.

Environmental Protection Department (2022) Emission factors for greenhouse gas inventories. *EPA*.

EPPM (2012) Environmental protection planning of Macao (2010–2020) (EN).

European Commission (2021a) A European green deal.

European Commission (2021b) Emissions cap and allowances.

Fairley, P. (2016) Why Southern China broke up its power grid. *IEEE*.

Fikri, M., *et al.* (2018) Estimating capital cost of small scale LNG carrier, pp. 225–229. doi: 10.5220/0008542102250229.

Fortune Oil (2020) Core businesses. *Fortune Oil*.

GEM (2023a) Global gas plant tracker. *Global Energy Monitor*.

GEM (2023b) Global solar power tracker. *Global Energy Monitor*.

George, Z., *et al.* (2020) LNG terminals in China – Project development, third party access and foreign investment issues.

Gerboni, R. (2016) Introduction to hydrogen transportation. *Compendium of Hydrogen Energy*, pp. 283–299. doi: 10.1016/b978-1-78242-362-1.00011-0.

Global CCS Institute (2012) Carbon dioxide distribution infrastructure.

Global CCS Institute (2020a) Global status of CCS: Brief for policymakers.

Global CCS Institute (2020b) Global status of CCS report. *Global CCS Institute*.

Global CCS Institute (GCI) (2020c) Global status of CCS. *Global CCS Institute*.

Global CCS Institute (2022) Global status of CCS 2022. *Global CCS Institute*.

Global Energy Monitor (2020a) Fossil tracker.

Global Energy Monitor (2020b) Global coal plant tracker. *Global Energy Monitor*.

Global Energy Monitor (2023) Global steel plant tracker. *GEM*.

Global Energy Observator (2011) *Global Energy Observator*.

Global Offshore Wind (2023) 4C Offshore.

Gondal, I. A. (2016) Hydrogen transportation by pipelines. *Compendium of Hydrogen Energy*, pp. 301–322. doi: 10.1016/b978-1-78242-362-1.00012-2.

Government of Macao Special Administrative Region Statistics and Census Service (2020) *Statistics Database*.

Gray, A. (2018) Shenzhen just made all its buses electric taxis are next. *World Economic Forum*.

Gray, M. (2020) Here's why China's post-COVID-19 stumulus must reject costly coal power. *Carbon Tracker*.

Greater Bay Insight (2019) BYD Changes Shenzhen next the world. *Greater Bay Insight*.

Grid Integration Group (2020) Microgrids and vehicle-grid integration. *Grid Integration Group*.

Guangdong Bureau of Statistics and National Bureau of Statistics (2020) *Guangdong Statistical Yearbook 2019*.

Guangdong Bureau of Statistics and National Bureau of Statistics (2022) *Guangdong Statistical Yearbook 2022*. China Statistics Press.

Guangdong Energy Group (2019) 2019 Guangdon Energy Group Annual Report.
Guangdong Energy Group (2020) Guangdong Energy Annual Report 2019.
Guangdong Provincial Government (2022) The 14th five-year plan of energy development in Guangdong Province. http://www.gd.gov.cn/attachment/0/486/486725/3909371.pdf.
Guardian, T. (2021) US-China deal on emissions welcomed by global figures and climate experts.
Gunia, A. (2021) Why China's promise to stop funding coal plants around the world is a really big deal. *Times*.
Guy, C. (2005) Air emissions from waste-to-energy plants.
Handa, R. and Baksi, S. (2020) CCUS: A climate-friendly approach to India's $5 trillion economy. *Down to Earth*.
Harrison, C. and Muething, L. (2022) Sustainable debt global state of the market 2021. *Climate Bonds Initiative*.
Hart, C., Zhu, J., and Ying, J. (2018) China_Climate_Map_Public_Secured_2019-3-1.
Hart, M., Bassett, L., and Johnson, B. (2017) Everything you think you know about coal in China is wrong. *Center for American Progress*.
Helfre, J.-F. and Depetiteville, E. (2022) Measuring the impact of green bonds. *S&P Global*.
Hellenic News (2020) China set to expand commercial crude storage capacity by 15.11 mil cu m in 2020. *Hellenic News*.
HK Census and Statistics Department (2022) Hong Kong Energy Statistics 2021 Annual Report. Hong Kong.
HK Electric (2014) The power behind Hong Kong.
HK Electric (2020a) Annual Report 2019. *HK Electric*. doi: 10.3934/math.2020i.
HK Electric (2020b) HKEI Sustainability Report.
HKEI (2021) HKEI 2020 Annual Report.
HKEX (2022) HKEX launches core climate, Hong Kong's international carbon marketplace, supporting global transition to net zero. *HKEX*.
HKGOV (2015) 2015 ~ 2025 + Environment Bureau in collaboration with Development Bureau Transport and Housing Bureau Message from the Chief Executive Special Messages from Principal Officials.

HKGOV (2017a) Hong Kong: The facts — Water, power and gas supplies, pp. 1–2.
HKGOV (2017b) Hong Kong's climate action plan 2030+.
HKGOV (2017c) I-Clean Air Plan for Hong Kong 2013–2017 Progress Report.
HKGOV (2018) Hong Kong: The facts power and gas supplies.
HKGOV (2019a) Hong Kong 2018 energy statistics.
HKGOV (2019b) Scheme of control agreements. *HKGOV*.
HKGOV (2020a) Figure 1: Revenue collected by tax type.
HKGOV (2020b) Greenhouse gas emissions in Hong Kong. doi: 10.1016/S1352-2310(00)00115-1.
HKGOV (2020c) Hong Kong energy statistics 2019.
HKGOV (2020d) The Chief Executive's 2020 policy address.
HKGOV (2021a) Climate action plan 2050 (October).
HKGOV (2021b) Hong Kong backs sustainable development with ESG focus. *Forbes*.
HKGOV (2021c) LCQ15: Employment market.
HKGOV (2021d) Roadmap on popularisation of electric vehicles.
HKGOV (2021e) Waste blueprint for Hong Kong 2035. doi: 10.32964/tj20.2.
HKGOV LEGCO (2021) Usage of electric vehicles in Hong Kong, (1), 20–21.
HKIMR (2020) The green bond market in Hong Kong: Developing a robust ecosystem for sustainable growth.
Hongkong Electric (2002) Corporate information. *Reference Reviews*, 16(2), 21–21. doi: 10.1108/rr.2002.16.2.21.77.
Hongkong Electric (2019) 2019–2023 Development Plan & 2019 Tariff Review Annex HEC-A Provision of information by The Hongkong Electric Company, Limited (HEC) for Economic Development Panel of the Legislative Council Information related to the 2019–2023 Development Plan Capital.
Hongkong Electric (2022) Corporate information.
Hove, A. and Sandalow, D. (2019) Electric vehicle charging in China and the United States. *Columbia Centre of Global Energy Policy* (February), pp. 1–86.
Hsieh, L., Pan, S., and Green, W. (2020) Transition to electric vehicles in China: Implications for private motorization rate and battery market. *Energy Policy*, 144.

Huang, W., *et al.* (2020) SWOT analysis of offshore wind power development in Guangdong Province. *IOP Conference Series: Earth and Environmental Science*, 510(2). doi: 10.1088/1755-1315/510/2/022040.

Hydrogen Council (2020) Path to hydrogen competitiveness: A cost perspective (January), p. 88.

ICAP (2020) China-Guangdong pilot ETS general information.

ICAP (2022) ICAP allowance price explorer. *ICAP*.

IEA (2019) Will energy from waste become the key form of bioenergy in Asia? *IEA*.

IEA (2020a) Carbon capture, utilisation and storage. *IEA*.

IEA (2020b) CCUS in clean energy transitions.

IEA (2020c) China's emissions trading scheme.

IEA (2020d) Coal 2020. *IEA*.

IEA (2020e) *Energy Technology Perspectives 2020 — Special Report on Carbon Capture Utilisation and Storage*. doi: 10.1787/208b66f4-en.

IEA (2020f) Geothermal 2019 Annual Report (August 2019).

IEA (2021a) Global Hydrogen Review 2021. doi: 10.1787/39351842-en.

IEA (2021b) Renewables. Paris.

IEA (2022a) Energy efficiency: Energy efficiency. *International Energy Association.*

IEA (2022b) Global EV Outlook 2022 — Securing Supplies for an Electric Future. *Global EV Outlook 2022*, p. 221.

IEA (2022c) World Energy Outlook 2022.

IEAGHG (2022) IEAGHG 2020 summer school. *2020 Summer School*.

IGU (2020) 2020 World LNG Report, p. 68.

IHA (2020) 2020_hydropower_status_report (3). *International Hydropower Association.*

Industry About (2020) China industrial map. *Industry About*.

International Carbon Action Partnership (2021) China national ETS (January), pp. 1–5.

IRENA (2019) Geothermal energy.

IRENA (2020) Global Renewables Outlook Edition: 2020.

Jiang, X., *et al.* (2020) Pathways to net zero carbon emissions by 2050.

Jianjun, K. (2013) How to manage the Chinese coal value chain. *Carnegie Energy and Climate Program* (August).

Josh, D. (2019) United Kingdom without evaluation of the climate change impacts of waste incineration in the United Kingdom (October 2018).

Kamalinejad, M., Sheykhbahaee, A., and Mazaheri, S. (2016) Financial feasibility study between purchasing and hiring LNG carrier in Iranian LNG industry. *International Journal of Coastal & Offshore Engineering*, 1(1), 25–31.

Kjärstad, J., *et al.* (2016) Ship transport — A low cost and low risk CO_2 transport option in the Nordic countries. *International Journal of Greenhouse Gas Control* (Elsevier Ltd), 168–184. doi: 10.1016/j.ijggc.2016.08.024.

Knoema (2017) Global GHG and CO_2 emissions.

Kong, A. A.-H. (2021) Aviation logistics services.

Land Department (2022) Land utilization in Hong Kong.

Landry, D. A., Thomas, M., and Becker, B. K. (2021) Carbon capture and sequestration in 2021: The path forward. *Liskow & Lewis.*

Li, T., Li, M., and Tian, L. (2021) Dynamics of carbon storage and its drivers in Guangdong Province from 1979 to 2012.

Liang, C. and Ang, S. (2021) Commodities 2021: China's natural gas demand set to hit new record. *S&P Global.*

Lin, C., *et al.* (2020) Assessing the impacts of large-scale offshore wind power in Southern China. *Energy Conversion and Economics*, 1(1), 58–70. doi: 10.1049/enc2.12006.

Liu, B., *et al.* (2018) Hydropower curtailment in Yunnan Province, southwestern China: Constraint analysis and suggestions. *Renewable Energy*, 121, 700–711. doi: 10.1016/j.renene.2018.01.090.

Liu, X. and Jin, Z. (2020) An analysis of the interactions between electricity, fossil fuels and carbon market prices in Guangdong China. *Energy for Sustainable Development*, 55, 82–94.

Liu, G., *et al.* (2022) China's pathways of CO_2 capture, utilization and storage under carbon neutrality vision 2060. *Carbon Management*, 435–449. doi: 10.1080/17583004.2022.2117648.

Lumiao, L. and Zhanhui, Y. (2020) New energy buses in China: Overview on policies and impacts — sustainable transport in China.

Lund, J. W. and Boyd, T. L. (2015) Direct utilization of geothermal energy 2015 worldwide review (April), 19–25.

Lydia, P., Browning, J., Aitken, A., Inman, M., and Nace, T. (2020) Gas bubble.

Meara, S. (2020) China's plan to cut coal and boost green growth. *Nature*, 584, S1–S3.

Microgrid Knowledge (2021) EVs are a disruptive force. Smart microgrids are the answer to manage the disruption. *Microgrid Knowledge*.

Miller, G. (2021) New world record set for shipping rates: $350,000 per day. *Freight Waves*.

Mirae Asset (2020) Guangdong to close coal power plants.

Mohindru, S. (2020) Oil tankers face new normal amid pandemic and decarbonization drive. *S&P Global*.

Monteith, S. and Wang, T. (2020) China pledged net-zero emissions by 2060. Here's what it will take to get there. *Climate Works Foundation*.

Moran, D., *et al.* (2018) Carbon footprints of 13 000 cities. *Environmental Research Letters*, 13(6). doi: 10.1088/1748-9326/aac72a.

Morton, A. (2021) World's biggest battery with 1,200MW capacity set to be built in NSW Hunter Valley. *The Guardian*.

MTMM (Hong Kong) (2020) Tanker berths. *MTMM (Hong Kong)*.

Mudie, L. (2020) Guangdong cities in temporary darkness amid China power glitch. *Radio Free Asia*.

Mukherjee, A. (2022) BYD widens gap with Tesla in Q3 2022, leads global EV market. *Counterpoint*.

Mullen, A. (2021) China coal: Why is it so important to the economy? *South China Morning Post*.

MWR (2016) Dam construction and management construction and management in China.

Myllyvirta, L. and Yedan, L. (2020) China's Covid stimulus plans for fossil fuels three times larger than low-carbon. *Carbon Brief*.

Myllyvirta, L., Zhang, S., and Shen, X. (2020) Analysis: Will China build hundreds of new coal plants in the 2020s? *Carbon Brief*.

Narayanan, A. (2020) Electric car sales more than double in worlds biggest EV market, but Tesla lags. *Investor Business Daily*.

Nash, D., *et al.* (2012) Hydrogen storage: Compressed gas. *Comprehensive Renewable Energy*, 4, 131–155. doi: 10.1016/B978-0-08-087872-0.00413-3.

National Bureau of Statistics (2020a) Chapter 7: Energy data for Hong Kong and Macao special administrative region. *China Energy Yearbook*.

National Bureau of Statistics (2020b) Industrial production operation in November 2020. *National Bureau of Statistics*.

National Energy Administration (2019) *2019*年上半年光伏发电建设运行情况.

National People's Congress (2016) The 13 five-year plan for economic and social development.

NDC (2021) China's achievements, new goals and new measures for nationally determined contributions 1.

Nedopil, C. and De Boer, D. (2020) China must boost green finance to achieve carbon neutrality by 2060. *China Dialogue*.

Noh, H., *et al.* (2019) Conceptualization of CO_2 terminal for offshore CCS using system engineering process. *Energies*, 12(22). doi: 10.3390/en12224350.

NS Energy (2020a) Shenzhen East waste-to-energy plant. *NS Energy*.

NS Energy (2020b) Zhongke refinery and petrochemical project. *NS Energy*.

Our World in Data (2020) Coal production, our world in data. https://ourworldindata.org/.

Parzen, M., *et al.* (2021) Beyond cost reduction: Improving the value of energy storage in electricity systems.

Patel, S. (2020) Xuzhou 3 shows the future of subcritical coal power is sublime. *Power*.

Paulson Institute (2015) Paulson papers on standards power play: China's ultra-high voltage technology and global standards Paulson papers on standards about the authors (April).

Pei, J., *et al.* (2018) Spatial-temporal dynamics of carbon emissions and carbon sinks in economically developed areas of China: A case study of Guangdong Province. *Scientific Reports* (August), 1–15. doi: 10.1038/s41598-018-31733-7.

Power (2017) Who has the world's most efficient coal power plant fleet. *POWER*.

Power Technology (2020) Black point combined cycle power plant, Hong Kong. *Power Technology*.

Power Technology (2021) World's biggest hydroelectric power plants. *Power Technology*.

Qiao, Q. and Lee, H. (2019) The role of electric vehicles in decarbonizing China's transportation sector. *Environment and Natural Resources Program*.

Raju, T. B., *et al.* (2016) Study of volatility of new ship building prices in LNG shipping. *International Journal of e-Navigation and Maritime Economy*, 61–73. doi: 10.1016/j.enavi.2016.12.005.

REN 21 (2020) Renewables 2019 Status Report.

REN 21 (2021) Renewables 2021 Global Status Report.

Reuters (2010) Hong Kong oil terminals shut down ahead of typhoon. *Reuters*.

Reuters (2017) China's Guangzhou port says coal imports operating normally after storage halt.

Reuters (2018) China, Venezuela joint refinery complex slated for start-up in late 2021. *Caixin*.

Reuter (2020) Private firm Huaying to build 6 million T LNG terminal in southern China. *Reuters*.

Reuters (2021) Fertilizer makers ponder hydrogen production. *The Western Producer*.

Rioux, B., *et al.* (2016) Economic impacts of debottlenecking congestion in the Chinese coal supply chain. *Energy Economics*. doi: 10.1016/j.eneco.2016.10.013.

Ritchie, H. and Roser, M. (2017) *Our World in Data*.

Roldao, R. (2022) Carbon trading the Chinese way. *Energy Monitor*.

RTHK (2021) CLP plans to build offshore wind farm. *RTHK*.

Sanseverino, E. R., *et al.* (2021) Life-cycle land-use requirement for PV in Vietnam. *Energies*, 14(4). doi: 10.3390/en14040861.

Sargent and Lundy (2020) Capital cost and performance characteristic estimates for utility scale electric power generating technologies. *EIA*.

SCIO (2020) Sustainable development of transport in China. *The State Council Information Office of the People's Republic of China*.

Seabra, J. E. (2021) Biofuels in the global energy mix.

Selin, E. N. and Lehman, C. (2022) Biofeul. *Britannica*.

Service, R. F. (2018) Ammonia — A renewable fuel made from sun, air, and water — Could power the globe without carbon. *Science*. doi: 10.1126/science.aau7489.

Shakeel, F. M. and Malik, O. P. (2019) Vehicle-to-grid technology in a micro-grid using DC fast charging architecture. *2019 IEEE Canadian Conference of Electrical and Computer Engineering, CCECE 2019* (May 2019), pp. 1–4. doi: 10.1109/CCECE.2019.8861592.

Shan, Y., *et al.* (2022) City-level emission peak and drivers in China. *Science Bulletin*, 67(18). doi: 10.1016/j.scib.2022.08.024.

Shell (2019) Our business. *Shell*.

Shen, J. (2022) BYD reports 232% year-on-year increase in passenger EV deliveries after bumper 2021. *Technode*.

Siemens (2020) Siemens launches an advanced microgrid demonstration environment at its New Jersey R&D Hub. *Siemens*.

Simon, L. (2021) Five things you need to know about the Glasgow climate pact. *World Economic Forum*.

SINOPEC (2020) Sinopec 2019 Annual Report and Accounts.

SNAM (2020) Global Gas Report 2020.

Spencer, T., Berghmans, N., and Sartor, O. (2017) Coal transitions in China's power sector: A plant-level assessment of stranded assets and retirement pathway.

Staff, E. (editorial) (2020) Massive carbon capture growth in 2020 still isn't enough, report finds. *Engineering and Technology*.

State Council (2021) China sees record-high coal transport by rail in December 2020.

Statistics and Census Service (2023) Time series database. *Macau SAR Government*. https://www.dsec.gov.mo/ts/#!/step1/en-US.

Stern, B. A. and Berghout, N. (2021) Is carbon capture too expensive?

Sugden-Nalbandia, H. (2016) Operating ratio and cost of coal power generation.

Sun, L., *et al.* (2018) Assessment of CO_2 storage potential and carbon capture, utilization and storage prospect in China. *Journal of the Energy Institute*, 91(6), 970–977. doi: 10.1016/j.joei.2017.08.002.

Tan, J. (2020) A green wave of ESG is poised to break over China. *World Economic Forum*.

Tank Storage (2009) ExxonMobil to lease Hong Kong storage. *Tank Storage*.

Taylor, C. (2020) Hydrogen and CCS in future energy. *GEOEXPRO.*
The Oxford Institute for Energy Studies (2020) Natural gas in China's power sector: Challenges and the road ahead (December), pp. 1–15.
The Wind Power (2020a) China Guangdong. *The Wind Power.*
The Wind Power (2020b) Countries China offshore. *The Wind Power.*
The World Bank (2020) Supporting climate business opportunities in emerging markets.
The World Bank (2021) Carbon pricing dashboard.
Thornton, A. (2019) China is winning the electric vehicle race. *World Economic Forum.*
Tim, O. (2021) Navigating the pumped-storage development life cycle. *National Hydropower Association.*
Towngas (2019) Towngas gas production. *Towngas.*
Towngas (2020a) Towngas China Annual Report.
Towngas (2020b) Towngas Hong Kong 2019 Annual Report.
Towngas (2021a) ECO Environmental Investments Limited. *Towngas.*
Towngas (2021b) Towngas: Annual Report 2020.
Trading Economics (2023) EU carbon permits. https://tradingeconomics.com/commodity/carbon (Accessed: 13 October 2023).
Trondle, T. (2020) Supply-side options to reduce land requirements of fully renewable electricity in Europe. *PLoS One*, 15 (8 August). doi: 10.1371/journal.pone.0236958.
Tsoi, K. H., *et al.* (2022) Pioneers of electric mobility: Lessons about transport decarbonisation from two bay areas. *Journal of Cleaner Production*, 330, 129866. doi: 10.1016/j.jclepro.2021.129866.
UK GOV (2021) COP 26.
UNCTAD (2019) Review of Maritime Transport 2019.
UNCTAD (2020) Review of Maritime Transport 2020. *United Nations Conference on Trade and Development.*
UNCTAD (2021) Review of Maritime Report 2021. *United Nations Publications.*
UNEP (2022) COP15 ends with landmark biodiversity agreement. *UNEP.*
Varro, L. and Fengquan, A. (2020) China's net-zero ambitions: The next Five-Year Plan will be critical for an accelerated energy transition.

van de Ven, D. J., *et al.* (2021) The potential land requirements and related land use change emissions of solar energy. *Scientific Reports*, 11(1). doi: 10.1038/s41598-021-82042-5.

Vorrath, S. and Parkinson, G. (2020) Australia's Tesla big battery is no longer biggest battery in the world. *Renew Economy.*

Wang, C. and Ducruet, C. (2014) Transport corridors and regional balance in China: The case of coal trade and logistics To cite this version: HAL Id: halshs-01069149 Transport corridors and regional balance in China: The case of coal trade and logistics.

Wang, H., *et al.* (2019) Energy conversion of urban wastes in China: Insights into potentials and disparities of regional energy and environmental benefits. *Energy Conversion and Management*, 198 (August), 111897. doi: 10.1016/j.enconman.2019.111897.

Wang, J., Li, C., and Chen, D. (2020a) Energy 2021 China. *Energy 2021 9th Edition.*

Wang, P. T., *et al.* (2020b) Carbon capture and storage in China's power sector: Optimal planning under the 2°C constraint. *Applied Energy*, 263, 114694. doi: 10.1016/j.apenergy.2020.114694.

Wang, D., *et al.* (2021) Thermodynamic analysis of coal-fired power plant based on the feedwater heater drainage-air preheating system. *Applied Thermal Engineering*, 185 (May 2020), 116420. doi: 10.1016/j.applthermaleng.2020.116420.

Watson, F. and Ram, A. (2021) EU carbon prices crash ahead of December 2021 futures contract expiry. *S & P Global.*

Wei, W. (2016) Regional study on investment for transmission infrastructure in China based on the State Grid data (June). doi: 10.1007/s11707-016-0581-4.

Weigang, L. *et al.* (2020) Economic analysis of renewable energy in the electricity marketization framework: A case study in Guangdong, China. *Energy Front.*

Weiyuan, Y. (2020) Analysis of the development status of China's natural gas pipeline construction industry in 2020, the country accelerates the construction of natural gas pipelines.

White, V. (2019) World scale hydrogen production. *AirProducts.*

Wind, J. (2016) Hydrogen-fueled road automobiles – Passenger cars and buses. *Compendium of Hydrogen Energy*, 1st edn. Elsevier

Ltd. doi: 10.1016/b978-1-78242-364-5.00001-4. Woodhead Publishing Series in Energy 2016, Pages 3–21.

World Economic Forum (2019) This Chinese megacity is building a giant waste-to-energy plant. *World Economic Forum.*

World Energy Council (2019) Innovation Insights Brief 2019.

World Nuclear Association (2020) Nuclear power in China. *World Nuclear Association.*

World Nuclear Association (2022) Nuclear power in the world today. *World Nuclear Association.*

Wrede, I. (2022) Germany revives dirty coal amid Russian gas war. *DW.*

Wu, S. and Yang, Z. (2020) Availability of public electric vehicle charging pile and development of electric vehicle: Evidence from China. *Sustainability (Switzerland),* 12(16). doi: 10.3390/SU12166369.

Xu, M. and Maguire, G. (2020) China coal supplies to tighten this winter on import curbs, strong demand. *Reuters.*

Xu, M. and Stanway, D. (2021) China doubles new renewable capacity in 2020; still builds thermal plants. *Reuters.*

Xu, J., Yi, B., and Fan, Y. (2020) Economic viability and regulation effects of infrastructure investments for inter-regional electricity transmission and trade in China. *Energy Economics*, 91, 104890. doi: 10.1016/j.eneco.2020.104890.

Ying, H., *et al.* (2016) Study of a roadmap for carbon capture and storage development in Guangdong Province, China. *International Journal of Sustainable Energy*, 35(9).

Ylhe, X. (2022) China's $17 trillion carbon neutrality challenge. *Upstream.*

Yuanyuan, L. (2020) China installed more than 1000 EV charging stations per day in 2019. *Renewable Energy World.*

Zeng, X., Li, W., and Guo, X. (2021) The guangdong carbon emissions trading scheme: Progress, challenges and trends. https://www.efchina.org/Reports-en/report-lceg-20220427-en.

Zero Emissions Platform (2011) The costs of CO_2 capture, transport and storage: Post-demonstration CCS in the EU.

Zhang, L. and Bai, W. (2020) Risk assessment of China's natural gas importation: A supply chain perspective (237). doi: 10.1177/2158244020939912.

Zhang, D., *et al.* (2015) Waste-to-energy in China: Key challenges and opportunities. *Energies*, 8(12), 14182–14196. doi: 10.3390/en81212422.

Zhang, L., Chen, S., and Zhang, C. (2019) Geothermal power generation in China Status and prospects. *Energy Science & Engineering*.

Zhang, J., Sun, M., and Tan, Q. (2020) In review: Energy regulation in China. *Lexology*.

Zhang, Xiaohan, Winchester, N., and Zhang, Xilang (2017) The future of coal in China. *Energy Policy*.

Zhao, G. and Lawson, M. (2020a) LNG terminals in China – project development, third party access and foreign investment issues. *King&Wood Mallesons*.

Zhao, X. and Yutong, L. (2020b) China is about to run out of places to store crude oil. *Caixin*.

Zheng, J., Sun, X., and Zhou, Y. (2020) Electric passenger vehicles sales and carbon dioxide emission reduction potential in China's leading markets. *Cleaner Production*.

Zhou, H., *et al.* (2016) The China Southern Power Grid. *IEEE Power & Energy*.

Zhou, D., *et al.* (2018a) A long-term strategic plan of offshore CO_2 transport and storage in northern South China Sea for a low-carbon development in Guangdong Province, China. *International Journal of Greenhouse Gas Control*, 70, 76–87. doi: 10.1016/j.ijggc.2018.01.011.

Zhou, L., *et al.* (2018b) A comparative study on grid resource utilization rate between China Southern Power Grid and National Grid Plc of UK.

Zhou, Y., *et al.* (2018c) Emissions and low-carbon development in Guangdong-Hong Kong-Macao Greater Bay Area cities and their surroundings. *Applied Energy*, 228, 1683–1692. doi: 10.1016/j.apenergy.2018.07.038.

Zhu, Y., *et al.* (2020) Water transfer and losses embodied in the West–East electricity transmission project in China. *Applied Energy*, 275 (June), 115152. doi: 10.1016/j.apenergy.2020.115152.

Index

Milton Keynes UK
Ingram Content Group UK Ltd.
UKHW050927090724
444975UK00002BA/9

9 789811 290305